Feedback Linearization of RF Power Amplifiers

FEEDBACK LINEARIZATION OF RF POWER AMPLIFIERS

JOEL L. DAWSON
Stanford University

THOMAS H. LEE
Stanford University

Springer Science+Business Media, LLC

Library of Congress Cataloging-in-Publication

Title: Feedback Linearization of RF Power Amplifiers
Author (s): Joel Dawson and Thomas H. Lee
ISBN 978-1-4757-8880-8 ISBN 978-1-4020-8062-3 (eBook)
DOI 10.1007/978-1-4020-8062-3

Originally published by Kluwer Academic Publishers in 2004
Softcover reprint of the hardcover 1st edition 2004

Printed on acid-free paper.

To Mari

Contents

List of Figures

List of Tables

Acknowledgments

It is with great pleasure that we acknowledge the many people who have supported the work described in this book. In particular, Professor Stephen Boyd deserves credit for originally proposing the investigation of Chapter 2, and for working closely with us to bring it to fruition. Professors Bruce Wooley and Donald Cox graciously read a draft of this entire manuscript, and provided valuable and insightful comments.

We would also like to thank a number of institutions for their support of this investigation. Lucent Technologies, the National Science Foundation, and the Hertz Foundation all provided fellowship support, as did Stanford University through its Stanford Graduate Fellows program. National Semiconductor continues to provide Stanford students with free use of its 0.25μm CMOS process, an almost unbelievable luxury for students in our field. Agilent Technologies supported this work through the FMA program at CIS. This was largely due to the efforts of Dr. Jim Hollenhorst and Paul Corredoura, who in addition provided friendship and were sources of stimulating technical discussion.

Stanford's Center for Integrated Systems was a wonderful place to work, and this was due in large part to the presence of its graduate students. Dr. Dawson gladly acknowledges all members, past and present, of the Lee (SMIrC), Wooley, and Wong groups who have given their friendship and collaboration. Dr. David Su, formerly of the Wooley group, was particularly generous with his advice and insight during the hardware testing stages of this investigation.

Ann Guerra, the administrative assistant to Professors Lee and Wooley, has been a marvel at making administrative tasks run smoothly. She does this with a warmth, kindness, and humor that have greatly eased the passage of many students through the Ph.D. program. We take this opportunity to thank her for being a wonderful person to work with.

Dr. Dawson would also like to acknowledge his family, which was a source of unending love and support. They showed him that he is not, and never has been, alone in his endeavors. Finally, Marisol Negrón deserves a special acknowledgment for her steadfast love and support during the toughest days of this investigation. It is only fitting that this book be dedicated to her.

Chapter 1

INTRODUCTION

Research activity in the area of radio-frequency (RF) circuit design has surged in the last decade in direct response to the enormous market demand for inexpensive, portable, high data rate wireless transceivers. Our expectations for such transceivers, such as cellular phones, rise as they become seemingly ubiquitous. Once, the simple fact of a fairly reliable wireless voice connection was sufficient and even exciting. Now, crystal-clear voice with no lapses in coverage is actively sought, together with the capability to act as a web portal and even a digital assistant. All of this must be accomplished by a device that is cheap enough to be virtually given away, small enough to justify the claim of portability, and frugal enough with power demands to last a long time on a single battery charge.

Cellular phones are just one example of a market that has spurred recent research activity. Wireless local-area networks (WLAN's) are another relatively new application of RF circuit techniques, as is the popular Global Positioning System (GPS). Meeting this demand for a kind of general connectivity involves a host of fascinating technical challenges. Among these, many are associated with the power amplifier, the system block that drives the antenna in any radio transmitter.

1.1 Motivation

If the objective is an inexpensive, portable, high-performance transceiver, the desirability of certain circuit characteristics is clear. A low-cost solution is likely to be one in which as many circuit blocks as possible are implemented on the same chip: the cost savings result from the simplified PC (printed circuit) board. An inexpensive IC (integrated circuit) process, such as CMOS, translates directly into a cost savings. Portability implies at least two things from a circuit standpoint: small size, which is another advantage of a highly integrated solution, and a long battery lifetime. Long battery lifetimes motivate low-power circuit techniques, so we add low power dissipation to the growing list of design constraints. What is meant by "high-performance" depends on the context. For purposes of this book, high-performance implies the ability to communicate at the highest data rate possible for a given channel bandwidth. Achieving this goal directs the system designer to linear modulation techniques, and the circuit designer to a means of achieving high linearity in the transmitter.

A transceiver's performance according to the metrics of degree of integration, power consumption, and transmitter linearity is usually dominated by the performance of the power amplifier. At even modest output powers (a few hundred milliwatts) it is far and away the most power-hungry system block in a transceiver, and the large voltage swings at its output push it deep into nonlinear regions of operation. The devices in most IC processes impose a maximum usable DC power supply voltage. Further, high-Q impedance transformations, which cannot always be realized on-chip, are sometimes necessary to achieve high output power levels.

It follows that improvements in transmitter performance depend on the progress made with the power amplifier. That observation motivates the investigation described in this book.

1.2 Organization

The arc of this text generally proceeds from the abstract to the applied, culminating with a description of a fabricated chip designed to tie together many of the concepts treated here. Chapter 2 deals with the theoretical problem of realizing amplification with a cascade of stages. One design option is to employ *local* feedback around each of the individual stages. This chapter details the surprising result that a wide range of specifications, including linearity, can be optimized through intelligent choice of the feedback gains. That an optimum exists is perhaps not a surprise, but that this optimum can be found quickly and unambiguously is new and of considerable interest. The key is a technique called geometric programming.

The importance of linearity in radio transmitters is treated briefly in Chapter 3, together with a description of the tradeoff between linearity and power efficiency in power amplifiers. This chapter is also an exploration of the various common methods of softening this tradeoff, which can be grouped under the general heading of "linearization techniques."

Chapter 4 describes a new approach for achieving and maintaining phase alignment in Cartesian feedback power amplifiers. The focus here is on the theoretical principles of the new method, which during the investigation were validated by simulation and by a discrete-component prototype. A new technique for realizing accurate analog multiplication is developed as a means of improving the performance of the first prototype. A full analysis of this multiplication technique is presented here.

This book concludes with Chapter 5, a description of the culminating IC prototype, and Chapter 6, final thoughts. Readers interested in further details of the hardware prototypes are directed to the appendices.

Chapter 2

OPTIMAL ALLOCATION OF LOCAL FEEDBACK IN MULTISTAGE AMPLIFIERS

The use of linear feedback around an amplifier stage was pioneered by Black [1], Bode [2], and others. The relations among the choice of feedback gain and the (closed-loop) gain, bandwidth, rise-time, sensitivity, noise, and distortion properties, are well understood (see, *e.g.*, [3]). For a single stage amplifier, the choice of the (single) feedback gain is a simple problem.

In this chapter we consider the *multistage* amplifier shown in figure 2.1, consisting of n open-loop amplifier stages denoted $A_1, \dots, A_n$, with local feedback gains $f_1, \dots, f_n$ employed around the stages.[1]

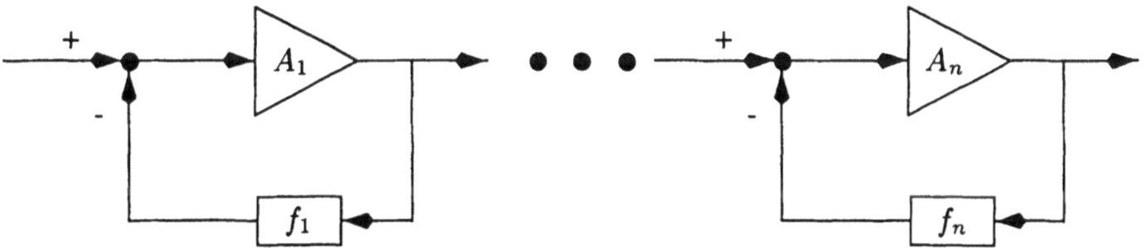

Figure 2.1. Block diagram of multistage amplifier.

We assume that the amplifier stages are fixed, and consider the problem of choosing the feedback gains $f_1, \dots, f_n$. The choice of these feedback gains affects a wide variety of performance measures for the overall amplifier, including gain, bandwidth, rise-time, delay, noise, distortion and sensitivity properties, maximum output swing, and dynamic range. These performance measures depend on the feedback gains in a complicated and nonlinear manner. It is thus

[1]Much of the material presented in this chapter originally appeared in the journal article [4], written by the author and coauthored by S. Boyd, M. Hershenson, and T. H. Lee.

far from clear, given a set of specifications, how to find an optimal choice of feedback gains. We refer to the problem of determining optimal values of the feedback gains, for a given set of specifications on overall amplifier performance, as the *local feedback allocation problem.*

We will show that the local feedback allocation problem can be cast as a *geometric program* (GP), which is a special type of optimization problem. Even complicated geometric programs can be solved very efficiently, and globally, by recently developed interior-point methods (see [5, 6, 7]). Therefore we are able to give a complete, global, and efficient solution to the local feedback allocation problem.

In section 2.1, we give a detailed description of the models of an amplifier stage used to analyze the performance of the amplifier. Though simple, the models capture the basic qualitative behavior of a source-degenerated differential pair. In section 2.2, we derive expressions for the various performance measures for the overall amplifier, in terms of the local feedback gains. In section 2.3, we give a brief description of geometric programming, and in section 2.4, we put it all together to show how the optimal local feedback allocation problem can be cast as a geometric program, and design examples are given in section 2.5. A summary of the method follows in section 2.6, along with a treatment of a specific circuit example in section 2.7. This chapter closes with section 2.8, a discussion of the relevance of local feedback allocation to power amplifier linearization.

2.1 Amplifier stage models

In this section we describe several different models of an amplifier stage, used for various types of analysis.

2.1.1 Linearized static model

The simplest model we use is the linear static model shown in figure 2.2. The stage is characterized by $y_i = \alpha_i e_i$, where α_i is the gain of the ith stage, which we assume to be positive. We will use this simple model for determining the overall gain of the amplifier, determining the maximum signal swing, and the sensitivity of the amplifier gain to each stage gain.

2.1.2 Static nonlinear model

To quantify nonlinear distortion effects, we use a static nonlinear model of the amplifier stage as shown in figure 2.3. We assume that the nonlinearity or transfer characteristic has the form

$$y_i = a_i(e) = \alpha_i e - \beta_i e^3 + o(e^3), \tag{2.1}$$

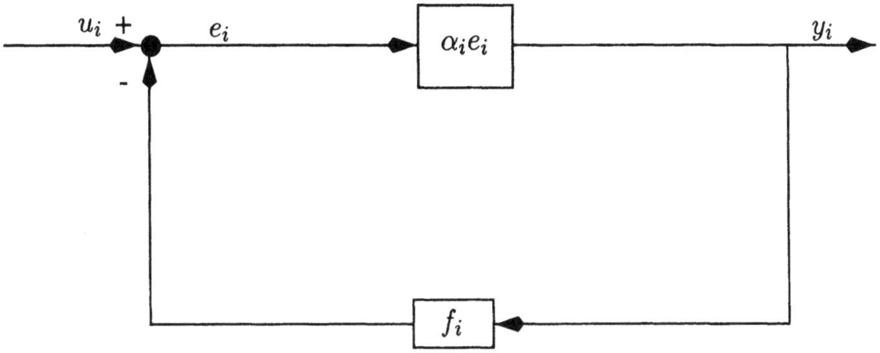

Figure 2.2. Linearized static model of amplifier stage.

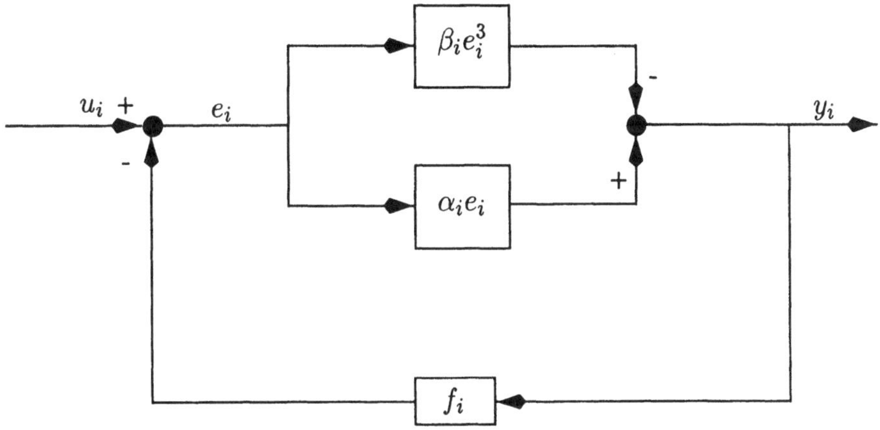

Figure 2.3. Nonlinear static model of amplifier stage.

where $o(e^3)$, which indicates terms of order higher than three, is assumed to be negligible. This form is inspired by the transfer characteristic of a source-coupled pair [8], and is a general model for third-order nonlinearity in a stage with an odd transfer characteristic. The function $a_i(\cdot)$ is the *transfer characteristic* of the ith stage, and β_i is the *third-order coefficient* of the amplifier stage. Note that the gain and third-order coefficient are related to the transfer characteristic by

$$\alpha_i = a_i'(0), \quad \beta_i = -\frac{a_i'''(0)}{6}. \tag{2.2}$$

We assume that $\beta_i \geq 0$, which means the third-order term is *compressive*: as the signal level increases from zero, the nonlinear term tends to decrease the output amplitude when compared to the linear model.

2.1.3 Linearized dynamic model

To characterize the bandwidth, delay, and rise-time of the overall amplifier, we use the linearized dynamic model shown in figure 2.4. Here the stage is represented by a simple one-pole transfer function with time constant τ_i (which we assume to be positive).

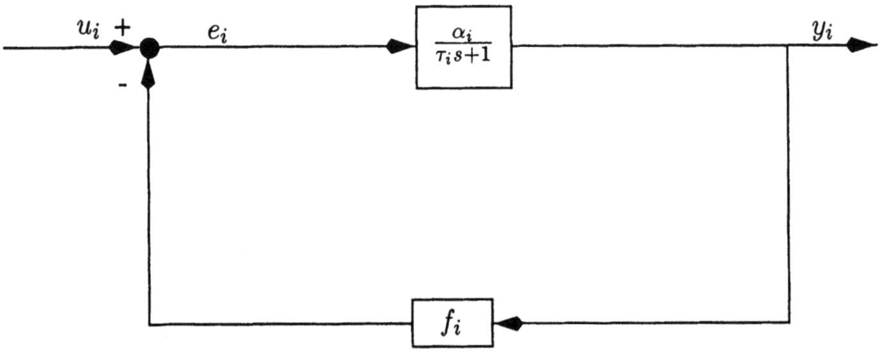

Figure 2.4. Linear dynamic model of amplifier stage.

2.1.4 Static noise model

Last, we have the static noise model shown in figure 2.5, which includes a simple output-referred noise v_i. As will become clear later, more complicated noise models including input noise, or noise injected in the feedback loop, are also readily handled by this method. Our noise model is characterized by the rms value of the noise source, which we denote $\overline{v_i}$. We assume that noise sources associated with different stages are uncorrelated.

2.2 Amplifier analysis

In this section we derive expressions for various performance indices for the overall amplifier. For analytical convenience, we express these indices in terms of the *return differences*:

$$l_i = 1 + f_i \alpha_i. \tag{2.3}$$

In the analysis that follows, it is assumed that the dynamic interaction between amplifier stages can be formulated as shown in section 2.7.3.

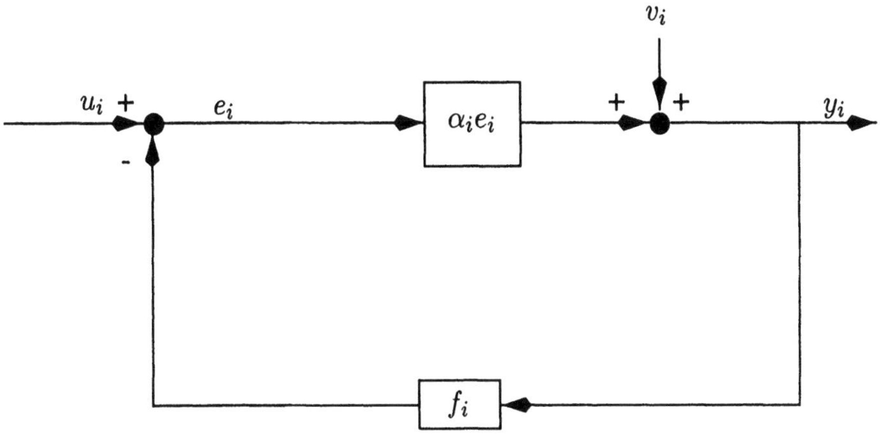

Figure 2.5. Static noise model of amplifier stage.

2.2.1 Gain and output swing

We consider the linear static model of section 2.1.1. The gain of the amplifier, from input u_1 to the output of the kth stage y_k, is given by

$$\prod_{i=1}^{k} \frac{\alpha_i}{1+\alpha_i f_i} = \prod_{i=1}^{k} \frac{\alpha_i}{l_i}, \tag{2.4}$$

and the overall gain, from u_1 to y_n, is given by

$$\tilde{\alpha} = \prod_{i=1}^{n} \frac{\alpha_i}{l_i}. \tag{2.5}$$

Here, of course, α_i/l_i is the familiar expression for the closed-loop gain of the ith stage. It will be convenient later to use the notation

$$\tilde{\alpha}_i = \frac{\alpha_i}{l_i} \tag{2.6}$$

to denote the closed-loop gain of the ith stage. (In general, we will use the tilde to denote a closed-loop quantity.)

Now suppose the input signal level is $U = |u_1|$, and that the ith stage has a maximum allowed output signal level of Y_i, *i.e.*, we require $|y_i| \leq Y_i$. This in turn means that for $k = 1, \ldots, n$, we have

$$U \prod_{i=1}^{k} \frac{\alpha_i}{l_i} \leq Y_k, \tag{2.7}$$

so the maximum allowed input signal level is

$$U_{\max} = \max_{k=1,\ldots,n} Y_k \prod_{i=1}^{k} \frac{l_i}{\alpha_i}. \tag{2.8}$$

The maximum allowed output signal level is found by multiplying by the overall gain:

$$Y_{\max} = \max_{k=1,\ldots,n} Y_k \prod_{i=k+1}^{n} \frac{\alpha_i}{l_i} \tag{2.9}$$

(where the empty product, when $i = n$, is interpreted as one).

2.2.2 Sensitivity

The (logarithmic) sensitivity of the overall amplifier gain to the open-loop gain of the ith stage is given by

$$S_i = \frac{\partial \log \tilde{\alpha}}{\partial \log \alpha_i} = \frac{1}{l_i}. \tag{2.10}$$

2.2.3 Nonlinearity

We begin by deriving the closed-loop third-order coefficient of a single feedback amplifier stage, using the static nonlinear model of section 2.1.2. The output y is related to the input u through the relation

$$y = a(u - fy). \tag{2.11}$$

Differentiating both sides with respect to u leads to the familiar result from elementary feedback theory:

$$y'(0) = \frac{a'(0)}{1 + fa'(0)} = \frac{\alpha}{l} = \tilde{\alpha}. \tag{2.12}$$

Differentiating again yields

$$y''(0) = \frac{a''(0)}{l^3} = 0, \tag{2.13}$$

and, once more,

$$y'''(0) = \frac{a'''(0)l - 3fa''(0)^2}{l^5} = -\frac{6\beta}{l^4}, \tag{2.14}$$

using $a'''(0) = -6\beta$ and $a''(0) = 0$ from the previous equation. This equation shows that the third-order coefficient of the closed-loop transfer characteristic is given by

$$\tilde{\beta} = \frac{y'''(0)}{6} = \frac{\beta}{l^4}. \tag{2.15}$$

This is the well-known result showing the linearizing effect of (linear) feedback on an amplifier stage.

Next, consider a cascade of two amplifier stages. Let the transfer characteristics of two stages be $y_1(\cdot)$ and $y_2(\cdot)$. We write

$$\phi_2(u) = y_2(y_1(u)) \tag{2.16}$$

and differentiate:

$$\begin{aligned} \phi_2'(0) &= y_2'(0)y_1'(0), \\ \phi_2''(0) &= y_2'(0)y_1''(0) + (y_1'(0))^2 y_2''(0), \end{aligned} \tag{2.17}$$

and so

$$\begin{aligned} \phi_2'''(0) = y_2'(0)y_1'''(0) + y_1'(0)^3 y_2'''(0) \\ +3y_2''(0)y_1'(0)y_1''(0). \end{aligned} \tag{2.18}$$

Since y_1 and y_2 are both odd functions, the last term vanishes. Therefore the third-order coefficient of the cascade of the two stages is given by

$$\tilde{\beta}_1\tilde{\alpha}_2 + \tilde{\beta}_2\tilde{\alpha}_1^3. \tag{2.19}$$

More generally, the third-order coefficient of a cascade of n stages can be expressed as

$$\tilde{\beta} = \sum_{i=1}^{n} \left[\left(\prod_{k=1}^{i-1} \tilde{\alpha}_k^3 \right) \tilde{\beta}_i \left(\prod_{j=i+1}^{n} \tilde{\alpha}_j \right) \right]. \tag{2.20}$$

This very complicated formula gives the relation between the local return differences and the third-order coefficient of the overall amplifier.

2.2.4 Bandwidth

We next examine the linearized dynamic performance of the amplifier chain, using the stage model given in section 2.1.3. The transfer function of an individual stage is given by

$$\frac{Y_i(s)}{U_i(s)} = \frac{\alpha_i/(\tau_i s + 1)}{1 + f_i\alpha_i/(\tau_i s + 1)} = \frac{\tilde{\alpha}_i}{\tilde{\tau}_i s + 1}, \tag{2.21}$$

where $\tilde{\tau}_i = \tau_i / l_i$ is the closed-loop time constant of the ith stage. The transfer function of the entire cascade amplifier immediately follows:

$$H(s) = \prod_{i=1}^{n} \frac{\tilde{\alpha}_i}{\tilde{\tau}_i s + 1}. \tag{2.22}$$

The -3dB bandwidth of the amplifier is defined as the smallest frequency ω for which $|H(j\omega)| = |H(0)|/\sqrt{2}$.

2.2.5 Delay and rise-time

The rise-time and delay of the overall amplifier can be characterized in terms of the moments of the impulse response, as described in [9]. The delay is the normalized first moment of the impulse response of the system:

$$t_{\mathrm{d}} = \frac{\int_{-\infty}^{\infty} t h(t) dt}{\int_{-\infty}^{\infty} h(t) dt}. \tag{2.23}$$

Using basic properties of the Laplace transform and results from section 2.2.4, we have

$$t_{\mathrm{d}} = \frac{-\left.\frac{dH(s)}{ds}\right|_{s=0}}{H(0)} = -\frac{d}{ds} \prod_{i=1}^{n} \left.\frac{1}{\tilde{\tau}_i s + 1}\right|_{s=0} = \sum_{i=1}^{n} \tilde{\tau}_i = \sum_{i=1}^{n} \frac{\tau_i}{l_i}. \tag{2.24}$$

This formula shows the exact relation between the overall amplifier delay (as characterized by the first moment of the impulse response) and the local return differences l_i.

We use the second moment of the impulse response,

$$t_{\mathrm{r}}^2 = 4 \left[\frac{\int_{-\infty}^{\infty} t^2 h(t) dt}{\int_{-\infty}^{\infty} h(t) dt} - t_{\mathrm{d}}^2 \right], \tag{2.25}$$

as a measure of the square of the rise-time of the overall amplifier in response to a step input. Again making use of Laplace transform identities, we express (2.25) in terms of the transfer function $H(s)$:

$$t_{\mathrm{r}}^2 = \frac{4}{H(0)} \left[\left. \frac{d^2}{ds^2} H(s) \right|_{s=0} - t_{\mathrm{d}}^2 \right]. \tag{2.26}$$

Substituting the transfer function of the amplifier, given in equation (2.22), we find that the rise-time of the overall amplifier is

$$t_{\mathrm{r}}^2 = 4 \sum_{i=1}^{n} \frac{\tau_i^2}{l_i^2} \tag{2.27}$$

(using the fact that the closed-loop rise-time of the ith stage is $2\tilde{\tau}_i$).

2.2.6 Noise and dynamic range

We now consider the static noise model of section 2.1.4. The mean-squared noise amplitude at the output of the overall amplifier can be written

$$\overline{v_{\mathrm{out}}^2} = \sum_{i=1}^{n} \left(\frac{\overline{v_i^2}}{l_i^2} \left[\prod_{j=i+1}^{n} \frac{\alpha_j}{l_j} \right]^2 \right). \tag{2.28}$$

The input-referred mean-squared noise is then

$$\overline{v_{\text{in}}}^2 = \frac{\overline{v_{\text{out}}}^2}{\tilde{\alpha}^2} = \sum_{i=1}^{n}\left(\frac{\overline{v_i}^2}{\alpha_i^2}\left[\prod_{j=1}^{i-1}\frac{l_j}{\alpha_j}\right]^2\right). \tag{2.29}$$

The *dynamic range* of the amplifier is the ratio of maximum output voltage to output-referred RMS noise level, expressed in decibels:

$$\text{DR} = 20\log_{10}\frac{Y_{\text{max}}}{\overline{v_{\text{out}}}}. \tag{2.30}$$

2.2.7 SFDR and IIP linearity measures

We conclude this analysis by obtaining expressions for the spurious-free dynamic range (SFDR) and the input-referred third-order intercept point (IIP3). They are both readily derived from the results in 2.2.3 through 2.2.6, and so contain no new information or analysis, but they are widely used performance indices for the amplifier.

SFDR and IIP3 give information about the linearity of an amplifier. They concern the results of the following experiment: inject a signal $v\cos\omega_1 t + v\cos\omega_2 t$ at the input, and examine the output for the presence of intermodulation (IM) products. We concern ourselves here with third-order IM products, which owe their existence to non-zero β_i. The third order intermodulation products are:

$$\left[\frac{3}{4}\tilde{\beta}v^3\right][\cos(\omega_1+2\omega_2)t + \cos(\omega_1-2\omega_2)t + \cos(2\omega_1+\omega_2)t + \cos(2\omega_1-\omega_2)t]. \tag{2.31}$$

The SFDR is defined as the signal-to-noise ratio when the power in each third order intermodulation product equals the noise power at the output [10]. To derive the SFDR, we simply refer a third order IM product back to the input and equate its amplitude to the input-referred RMS noise amplitude:

$$\frac{1}{\tilde{\alpha}}\left[\frac{3}{4}\tilde{\beta}v^3\right] = \overline{v_{\text{in}}}. \tag{2.32}$$

The SFDR in decibels is then given by

$$\text{SFDR} = \frac{2}{3}(10)\log_{10}\left(\frac{\frac{4}{3}\left|\frac{\tilde{\alpha}}{\tilde{\beta}}\right|}{\overline{v_{\text{in}}}^2}\right). \tag{2.33}$$

The IIP3 is the input power at which the amplitude of the third-order IM products equals the input. Mathematically, we require

$$|\tilde{\alpha}v| = \left|\frac{4}{3}\tilde{\beta}v^3\right|. \tag{2.34}$$

Normalizing the input resistance to unity for convenience, we have for IIP3

$$\mathrm{IIP3} = \frac{2}{3}\left|\frac{\tilde{\alpha}}{\tilde{\beta}}\right|. \tag{2.35}$$

2.3 Geometric programming

Let f be a real-valued function of n real, positive variables $x_1, x_2, \ldots, x_n$. It is called a *posynomial* function if it has the form

$$f(x_1, \ldots, x_n) = \sum_{k=1}^{t} c_k x_1^{\alpha_{1k}} x_2^{\alpha_{2k}} \cdots x_n^{\alpha_{nk}} \tag{2.36}$$

where $c_j \geq 0$ and $\alpha_{ij} \in \mathbf{R}$. When $t = 1$, f is called a *monomial* function. Thus, for example, $0.7 + 2x_1/x_3^2 + x_2^{0.3}$ is posynomial and $2.3(x_1/x_2)^{1.5}$ is a monomial. Posynomials are closed under sums, products, and nonnegative scaling.

A *geometric program* (GP) has the form

$$\begin{array}{ll} \text{minimize} & f_0(x) \\ \text{subject to} & f_i(x) \leq 1, \quad i = 1, 2, \ldots, m, \\ & g_i(x) = 1, \quad i = 1, 2, \ldots, p, \\ & x_i > 0, \quad i = 1, 2, \ldots, n, \end{array} \tag{2.37}$$

where f_i are posynomial functions and g_i are monomial functions. Geometric programs were introduced by Duffin, Peterson, and Zener in the 1960s [11].

The most important property of geometric programs for us is that they can be solved, with great efficiency, and globally, using recently developed interior-point methods [7, 5]. Geometric programming has recently been used to optimally design electronic circuits including CMOS op-amps [12, 13] and planar spiral inductors [14].

Several simple extensions are readily handled by geometric programming. If f is a posynomial and g is a monomial, then the constraint $f(x) \leq g(x)$ can be expressed as $f(x)/g(x) \leq 1$ (since f/g is posynomial). In particular, constraints of the form $f(x) \leq a$, where $a > 0$ is a constant, can also be used. Similarly, if g_1 and g_2 are both monomial functions, the constraint $g_1(x) = g_2(x)$ can be expressed as $g_1(x)/g_2(x) = 1$ (since g_1/g_2 is monomial). If g is a monomial, we can maximize it by minimizing the posynomial function $1/g$.

2.3.1 Geometric programming in convex form

A geometric program can be reformulated as a *convex optimization problem*, *i.e.*, the problem of minimizing a convex function subject to convex inequalities constraints and linear equality constraints. This is the key to our ability to globally and efficiently solve geometric programs. We define new variables

$y_i = \log x_i$, and take the logarithm of a posynomial f to get

$$h(y) = \log\left(f\left(e^{y_1}, \ldots, e^{y_n}\right)\right) = \log\left(\sum_{k}^{t} e^{a_k^T y + b_k}\right) \tag{2.38}$$

where $a_k^T = [\alpha_{1k} \cdots \alpha_{nk}]$ and $b_k = \log c_k$. It can be shown that h is a *convex* function of the new variable y: for all $y, z \in \mathbf{R}^n$ and $0 \leq \lambda \leq 1$ we have

$$h(\lambda y + (1-\lambda)z) \leq \lambda h(y) + (1-\lambda)h(z). \tag{2.39}$$

Note that if the posynomial f is a monomial, then the transformed function h is affine, *i.e.*, a linear function plus a constant.

We can convert the standard geometric program (2.37) into a convex program by expressing it as

$$\begin{array}{lll} \text{minimize} & \log f_0(e^{y_1}, \ldots, e^{y_n}) & \\ \text{subject to} & \log f_i(e^{y_1}, \ldots, e^{y_n}) \leq 0, & i = 1, \ldots, m \\ & \log g_i(e^{y_1}, \ldots, e^{y_n}) = 0, & i = 1, \ldots, p. \end{array} \tag{2.40}$$

This is the so-called *convex form* of the geometric program (2.37). Convex programs have several important characteristics. Chief among these is that convex programs are solvable using efficient interior-point methods. Additionally, there is a complete and useful duality, or sensitivity, theory for convex programs [5].

2.3.2 Solving geometric programs

Since Ecker's survey paper [6] there have been several important developments related to solving geometric programs in the exponential form. A huge improvement in computational efficiency was achieved in 1994, when Nesterov and Nemirovsky developed efficient interior-point algorithms to solve a variety of nonlinear optimization problems, including geometric programs [7]. Recently, Kortanek et al. have shown how the most sophisticated primal-dual interior-point methods used in linear programming can be extended to geometric programming, resulting in an algorithm approaching the efficiency of current interior-point linear programming solvers [15]. The algorithm they describe has the desirable feature of exploiting sparsity in the problem, *i.e.*, efficiently handling problems in which each variable appears in only a few constraints.

For our purposes, the most important feature of geometric programs is that they can be *globally* solved with great efficiency. Problems with hundreds of variables and thousands of constraints are readily handled, on a small workstation, in minutes. The problems we encounter in this chapter, which have a few tens of variables and fewer than 100 constraints, are easily solved in under one second.

Perhaps even more important than the great efficiency is the fact that algorithms for geometric programming always obtain the global minimum. Infeasibility is unambiguously detected: if the problem is infeasible, then the algorithm will determine this fact, and not just fail to find a feasible point. Another benefit of the global solution is that the initial starting point is irrelevant. We emphasize that the same global solution is found no matter what the initial starting point is.

These properties should be compared to general methods for nonlinear optimization, such as sequential quadratic programming, which only find *locally* optimal solutions, and cannot unambiguously determine infeasibility. As a result, the starting point for the optimization algorithm does have an affect on the final point found. Indeed, the simplest way to lower the risk of finding a local, instead of global, optimal solution, is to run the algorithm several times from different starting points. This heuristic only reduces the risk of finding a nonglobal solution. For geometric programming, in contrast, the risk is always exactly zero, since the global solution is always found regardless of the starting point.

2.4 Optimal local feedback allocation

We now make the following observation, based on the results of section 2.2: a wide variety of specifications for the performance indices of the overall amplifier can be expressed in a form compatible with geometric programming using the variables l_i. The startling implication is that optimal feedback allocation can be determined using geometric programming.

The true optimization variables are the feedback gains f_i, but we will use instead the return differences l_i, with the constraints $l_i \geq 1$ imposed to ensure that $f_i \geq 0$. Once we determine the optimal values for l_i, we can find the optimal feedback gains via

$$f_i = (l_i - 1)/\alpha_i. \qquad (2.41)$$

2.4.1 Closed-loop gain

The closed-loop gain $\tilde{\alpha}$ is given by the monomial expression (2.5). Therefore we can impose any type of constraint on the closed-loop gain. We can require it to equal a given value, for example, or specify a minimum or maximum value for the closed-loop gain. Each of these constraints can be handled by geometric programming.

2.4.2 Maximum signal swing

The maximum output signal swing is given by the expression (2.9). The constraint that the output swing exceed a minimum required value, *i.e.*, $Y_{\max} \geq$

γ, can be expressed as

$$Y_k \prod_{i=k+1}^{n} \frac{\alpha_i}{l_i} \geq \gamma, \quad k = 1, \dots, n. \tag{2.42}$$

Each of these inequalities is a monomial inequality, and hence can be handled by geometric programming. Note that we also allow the bound on signal swing, *i.e.*, γ, as a variable here.

2.4.3 Sensitivity

The sensitivity of the amplifier to the ith stage gain is given by the monomial expression (2.10). It follows that we can place an upper bound on the sensitivity (or, if we choose, a lower bound or equality constraint).

2.4.4 Bandwidth

Consider the constraint that the closed-loop -3dB bandwidth should exceed Ω. Since the magnitude of the transfer function of the amplifier is monotonically decreasing as a function of frequency, this is equivalent to imposing the constraint

$$|H(j\Omega)| \geq \frac{1}{\sqrt{2}} H(0), \tag{2.43}$$

which we can rewrite as

$$\frac{1}{|H(j\Omega)|^2} \leq \frac{2}{H(0)^2}. \tag{2.44}$$

Now using the expression for the transfer function,

$$H(j\omega) = \prod_{i=1}^{n} \frac{\alpha_i}{j\tau_i\omega + l_i}, \tag{2.45}$$

we can write the bandwidth constraint as

$$\left|\frac{1}{H(j\Omega)}\right|^2 = \prod_{i=1}^{n} \frac{(\tau_i\Omega)^2 + l_i^2}{\alpha_i^2} \leq 2 \prod_{i=1}^{n} \frac{l_i^2}{\alpha_i^2}. \tag{2.46}$$

This, in turn, we can express as

$$\prod_{i=1}^{n} \frac{(\tau_i\Omega)^2 + l_i^2}{l_i^2} \leq 2. \tag{2.47}$$

This is a complicated, but posynomial, inequality in the variables l_i, hence it can be handled by geometric programming. Note that we can even make the minimum -3dB bandwidth Ω a variable, and maximize it.

2.4.5 Noise and dynamic range

The expression (2.29) for the input-referred noise power, is a posynomial functions of the variables $l_1, \ldots, l_n$. Therefore we can impose a maximum on the input-referred noise level, using geometric programming.

The requirement that the dynamic range exceed some minimum allowed value $\mathrm{DR}_{\mathrm{min}}$, *i.e.*, $\mathrm{DR} \geq \mathrm{DR}_{\mathrm{min}}$, can be expressed as

$$\frac{\overline{v_{\mathrm{out}}^2}}{\gamma} \leq 10^{-\mathrm{DR}/10}, \tag{2.48}$$

where γ is the bound on signal swing defined in (2.42). Therefore, this constraint can be handled by geometric programming.

2.4.6 Delay and rise-time

As can be seen in equations (2.24) and (2.27), the expressions for delay and rise-time are posynomial functions of the return differences l_i. A maximum on each can thus be imposed.

2.4.7 Third-order distortion

The expression for third-order coefficient, given in (2.20), is a posynomial, so we can impose a maximum on the third-order coefficient.

2.4.8 SFDR and IIP3

Consider the constraint that the SFDR should exceed some minimum value γ. Using the expression (2.33), we can write this as

$$\frac{2}{3}(10)\log_{10}\left(\frac{\frac{4}{3}\left|\frac{\tilde{\alpha}}{\tilde{\beta}}\right|}{\overline{{v_{\mathrm{in}}}^2}}\right) \geq \gamma. \tag{2.49}$$

This can be written as

$$\frac{4}{3}\frac{\tilde{\alpha}}{\tilde{\beta}}\frac{1}{\overline{{v_{\mathrm{in}}}^2}} \geq 10^{3\gamma/20}. \tag{2.50}$$

This can be handled by geometric programming by writing it as

$$\frac{\tilde{\beta}}{\tilde{\alpha}}\overline{{v_{\mathrm{in}}}^2} \leq \frac{3}{4}10^{-3\gamma/20}. \tag{2.51}$$

2.5 Design Examples

The foregoing analysis shows that complicated problems of feedback allocation can be expressed as globally solved geometric programs. We can take as an objective any of the posynomial performance measures described above, and apply any combination of the constraints described above. We can also compute optimal trade-off curves by varying one of the specifications or constraints over a range, computing the optimal value of the objective for each value of the specification.

We provide in this section a few system-level examples. In the section 2.7, we demonstrate a circuit-level application using the common source-coupled pair.

2.5.1 Trade-offs among bandwidth, gain, and noise

In our first example we consider a three-stage amplifier, with all stages identical, with parameters

$$\alpha_i = 10, \qquad \tau_i = 10^{-6}\text{sec}, \qquad \overline{v_{n,i}} = 4.07\mu\text{V}. \tag{2.52}$$

The required closed-loop gain is 23.5dB. We maximize the bandwidth, subject to the equality constraint on closed-loop gain, and a maximum allowed value of input-referred noise.

Figure 2.6 shows the optimal bandwidth achieved, as a function of the maximum allowed input-referred noise. As it must, the optimal bandwidth increases as we relax (increase) the input-referred noise limit. Figure 2.7 shows the optimal values of the feedback gains as the input-referred noise limit varies.

These curves roughly identify two regions in the design space. In one, the noise constraint is so relaxed as to not be an issue. The program identifies the optimum bandwidth solution for the given gain, which is to place all of the closed loop poles in the same place. In the other, the tradeoff between bandwidth and noise is strong. Equation (2.29) shows that the noise contribution of $\overline{v_{n,1}}$ is independent of l_1, but the noise contributions of the following stages can be diminished by making l_1 (and therefore f_1) small. It follows that f_3 is the greatest of the feedback gains, followed by f_2 and f_1.

We can also examine the optimal trade-off between bandwidth and required DC gain. Here we impose the fixed limit on input-referred noise at 4.15×10^{-7} V rms, and maximize the bandwidth subject to a required closed-loop gain.

Figures 2.8 and 2.9 show the maximum attainable bandwidth and the optimal feedback gain allocation as a function of the required closed-loop gain. Again we see two regions in design space caused by the noise constraint.

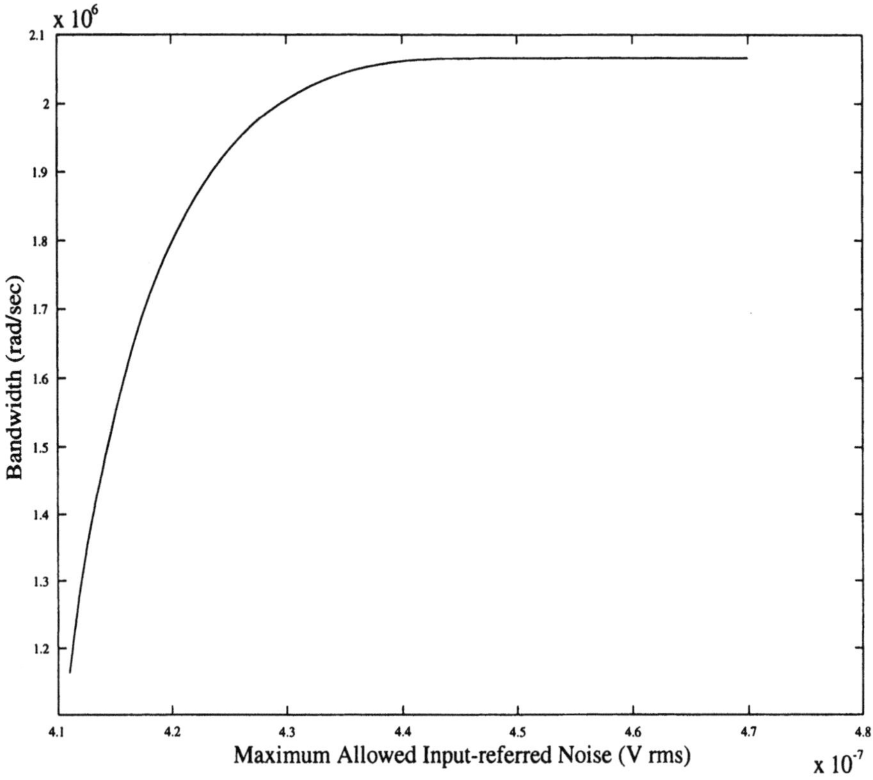

Figure 2.6. Maximum bandwidth versus limit on input-referred noise.

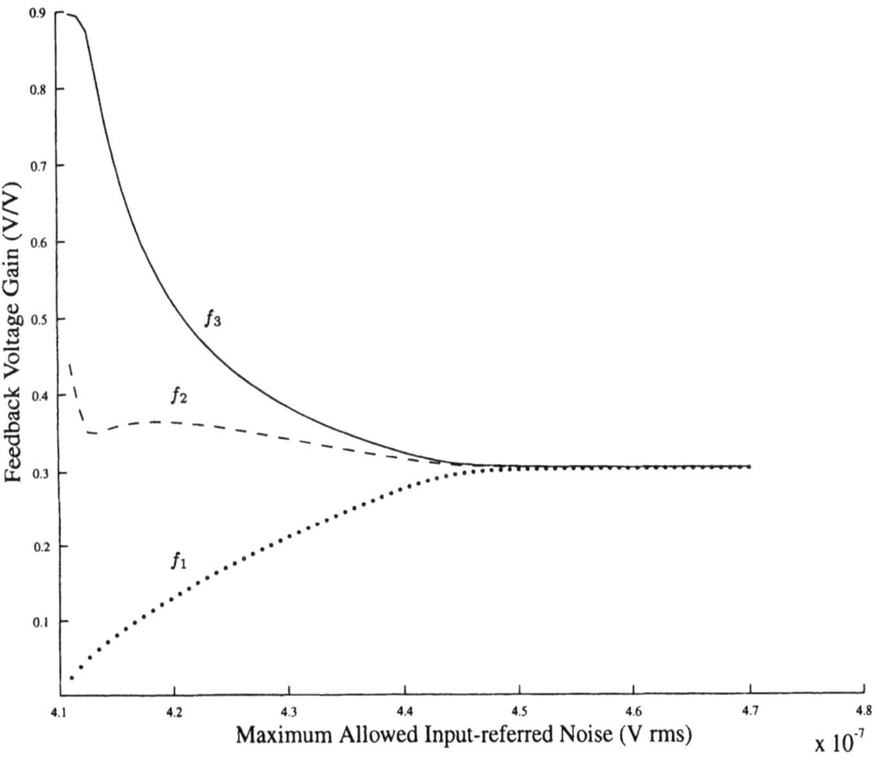

Figure 2.7. Optimal feedback allocation pattern, for maximum bandwidth with limit on input-referred noise. Gain = 23.5dB.

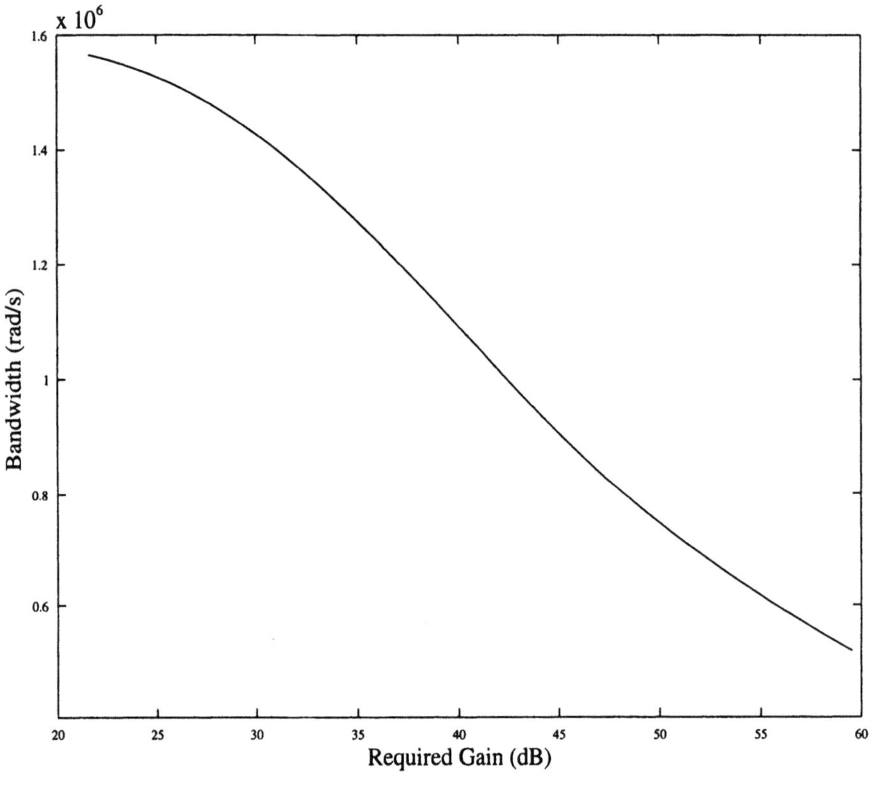

Figure 2.8. Maximum bandwidth versus required closed-loop gain. Maximum input-referred noise = 4.15e-7 V rms.

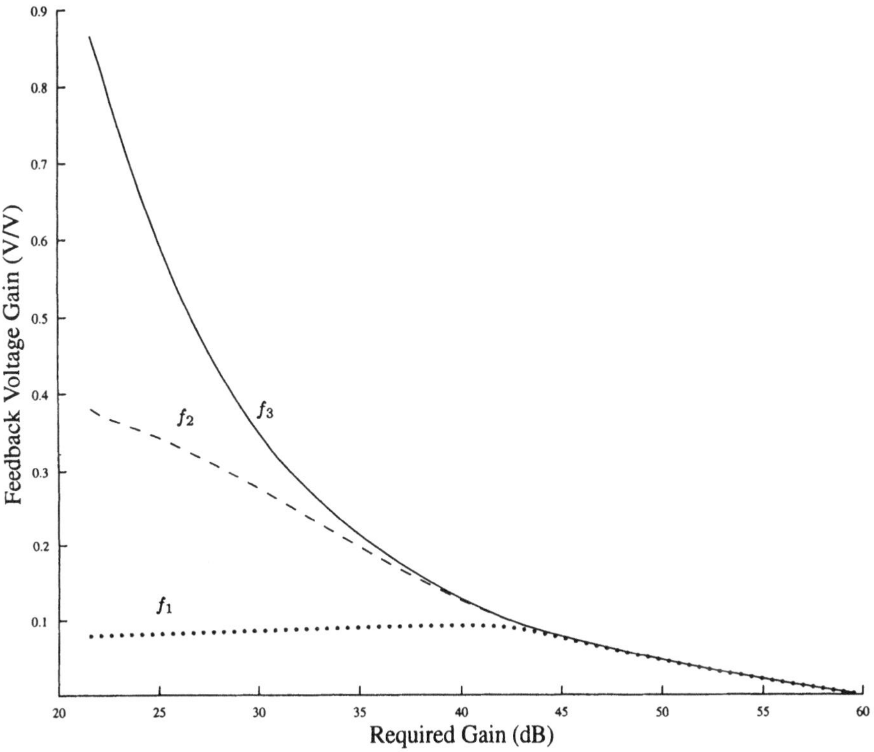

Figure 2.9. Optimal feedback allocation pattern for maximum bandwidth versus required closed-loop gain. Maximum input-referred noise = 4.15e-7 V rms.

2.5.2 SFDR versus gain

In this example, we again consider a three-stage amplifier, now with identical stages having parameters

$$\alpha_i = 10, \qquad \beta_i = 0.667\text{V}^{-2}, \qquad \overline{v_{n,i}} = 4.07\mu\text{V}. \tag{2.53}$$

We maximize the spurious-free dynamic range subject to an equality constraint on the overall gain. Figure 2.10 shows the achieved SFDR as a function of the required gain, and figure 2.11 shows the associated optimal gain allocation patterns.

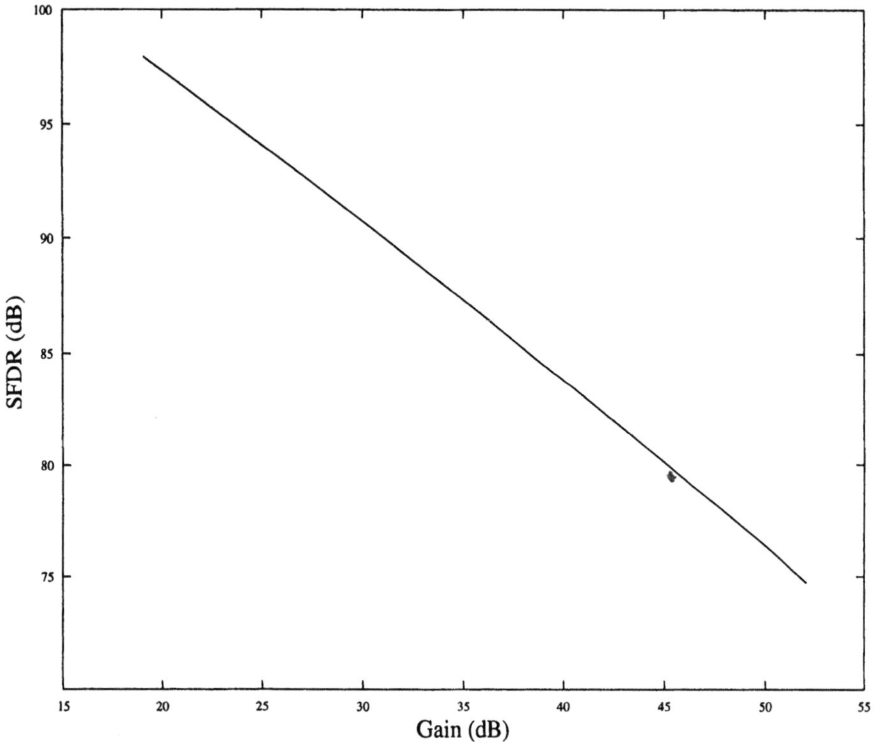

Figure 2.10. Maximum spurious-free dynamic range versus required gain.

In addition to obtaining optimal designs from the figures 2.10 and 2.11, we observe a qualitative trend: feedback gain is allocated preferentially to stages furthest down the signal chain. This is in agreement with sound engineering judgment, and with the results of section 2.5.1.

We can also argue from the standpoint of optimum linearity that figure 2.11 makes sense. Nonlinearity in the later stages, where the signal amplitude is the

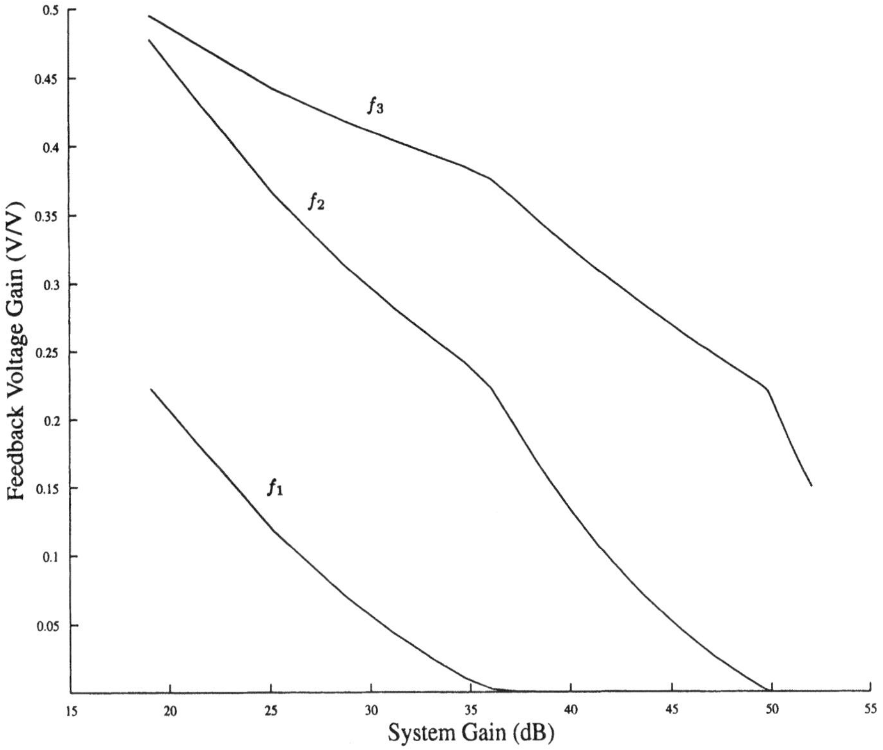

Figure 2.11. Optimal feedback allocation pattern for maximum spurious-free dynamic range versus required gain.

largest, will cause the most severe harmonic distortion. It follows that feedback should be applied more aggressively in later stages.

2.5.3 Stage selection

The method described in chapter computes the globally optimal values of the local feedback gains, with the amplifier models fixed. We can use the method indirectly to optimally choose each stage, from a set of possible choices, in addition to optimally allocating feedback around the stages. Suppose we have a set of m possible choices for each of n stages. By computing the optimal performance for each m^n possible combinations of stages, we can then determine the optimal combination as well as the optimal feedback gains. Of course the effort required to exhaustively search over the combinations grows rapidly with the number of stages, but is certainly feasible for moderate numbers of stages, *e.g.*, fewer than six or so.

We demonstrate this method for optimal stage selection with an example. Table 2.1 shows a listing of three candidate stages for use in a multistage amplifier design. Amplifier A can be seen to have the best linearity and the worst

Amplifier Designation	α_i	β_i	$\overline{v_n}$
A	10	0.67	$41\mu V$
B	10	6.7	$4.1\mu V$
C	10	67	$0.41\mu V$

Table 2.1. Candidate stages.

noise, amplifier C has the worst linearity and the best noise, and amplifier B is in between.

Our goal is to maximize SFDR, subject to a required gain of 46dB, for a three-stage design. By solving all 27 combinations, we find that the optimal combination of stages and feedback is

Stage 1 amplifier C, with f = 0.00023,

Stage 2 amplifier B, with f = 0.06589,

Stage 3 amplifier A, with f = 0.20072,

which achieves the optimal SFDR of 85dB.

This solution makes sense. The low-noise stage is used for the first stage (which is more critical for noise, since its noise is amplified by subsequent stages), and the high-linearity stage is used for the last stage (which handles larger signals, and so is more critical for distortion). Note that in this particular case, the optimal solution is to operate the first two stages essentially open-loop.

2.6 Geometric programming summary

In this chapter we show how to globally and efficiently solve the problem of optimally allocating local feedback gains in a multistage amplifier by posing the problem in the form of a geometric program. This formulation can handle a wide variety of practical objectives and constraints, and allows us to rapidly compute globally optimal trade-off curves among competing specifications.

We mention several extensions that are readily handled. While perhaps involved, it is not hard to work out the corresponding (posynomial) formulas for distortion characteristics that have fourth, fifth, or even higher order terms. It is also easy to handle a more sophisticated noise model, in which the noise is injected at several locations in the feedback around each stage, and not just at the output as in the current model. In each case, the resulting noise power expression is still posynomial, and therefore can be handled by geometric programming.

Another extension is to couple the design of the feedback together with the actual component-level design of the amplifier (for example, transistor widths and lengths), as in [16].

We envision several situations where the methods described in this chapter would be very useful to a circuit designer. Whenever the number of stages is at least three, and the number of important specifications is at least three (say), the problem of optimally allocating local feedback gains becomes quite complex, and a tool that completely automates this process is quite useful. When the number of stages reaches five or six, and the absolute optimal performance is sought, our method will far outperform even a good designer adjusting gains in an ad hoc manner.

2.7 An example application

The foregoing analysis establishes feedback allocation as a solvable problem. The extension of our technique to real world applications, however, begs clarification: we (seemingly) ignore loading between stages, choose suspiciously simple single-pole dynamics, etc.. We thus include the following example, in which we consider the ubiquitous source-coupled pair as our basic open-loop stage. Local feedback is allocated in the form of source degeneration, and all other characteristics (bias currents, load resistances, transistor sizes, etc.) are fixed.[2]

Figure 2.12 shows the basic stage that we consider. The differential half-

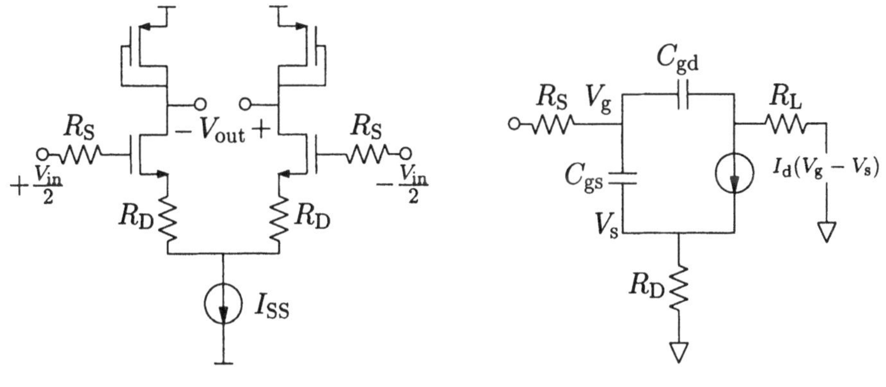

Figure 2.12. CMOS source-coupled pair and differential half-circuit.

circuit on the right should not be taken to represent the traditional "small-signal" model, as the dependent current source models a MOSFET operating in the saturation regime. The capacitors C_{gs} and C_{gd} *are* linear capacitors [10], and the

[2]These, too, can be optimized via geometric programming. See, for example, [13], [12].

PMOS devices provide the resistances R_L. We show in the following paragraphs how this common structure maps to the theoretical framework outlined in section 2.1.

2.7.1 Linearized static model

For this model the capacitors shown in figure 2.12 become open circuits, and the mapping from figure 2.2 to figure 2.13 is straightforward. A few short lines

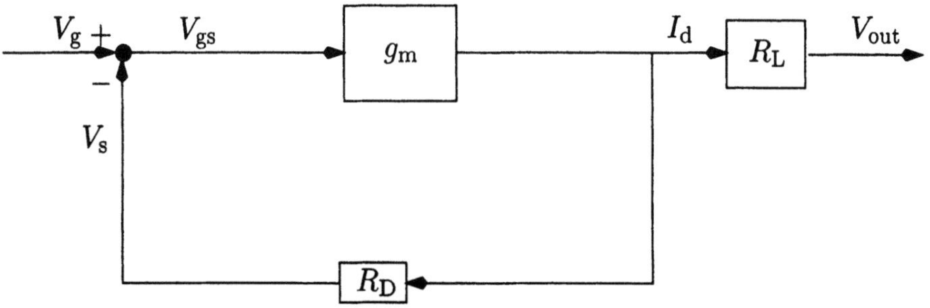

Figure 2.13. Source degeneration as a form of feedback.

of algebra lead to the familiar gain expression:

$$A_v = \left(\frac{g_m}{1 + g_m R_D}\right) R_L. \tag{2.54}$$

Already, it can be seen that l_i from the foregoing analysis finds its place here as $(1 + g_m R_D)_i$, with R_D taking the place of feedback gain. We emphasize that this is not merely a mathematical accident, but points to the physically meaningful interpretation of degeneration as a feedback mechanism.

2.7.2 Static nonlinear model

Here, we modify figure 2.13 by replacing g_m with $G_m(\cdot)$, the nonlinear expression for drain current as a function of V_{gs}. The expression for differential output current I_d as a function of differential gate voltage V_g for a general source-coupled pair is given in Gray and Meyer [8]. We reproduce it here:

$$I_d = \mu_n \frac{C_{ox} W}{2L} V_g \sqrt{\left(\frac{2 I_{SS}}{\mu_n \left(\frac{C_{ox} W}{2L}\right)}\right) - (V_g)^2}. \tag{2.55}$$

All constants in this formula are MOSFET parameters, and I_{SS} is the value of the current source in figure 2.12. For our purposes, a Taylor expansion of the square root allows us to write $G_m(\cdot)$ as

$$G_m(V_{gs}) = \alpha V_{gs} - \beta V_{gs}^3 + \gamma V_{gs}^5 - \cdots . \tag{2.56}$$

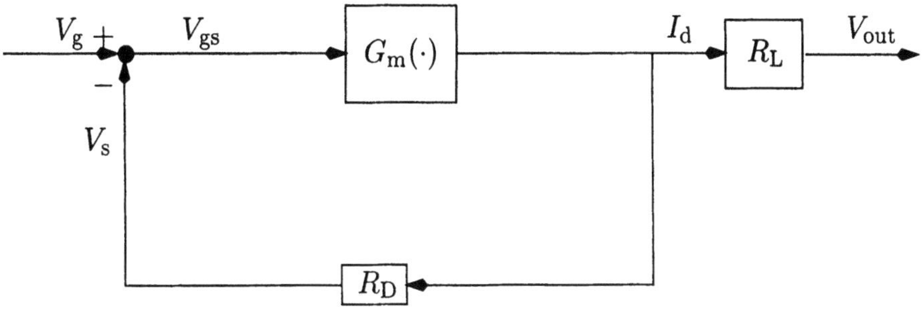

Figure 2.14. Modification for nonlinear static model.

This is consistent with figure 2.3.

2.7.3 Linearized dynamic model

We use the Miller approximation described in Gray and Meyer [8], modified here to account for source degeneration. The Miller approximation (see fig-

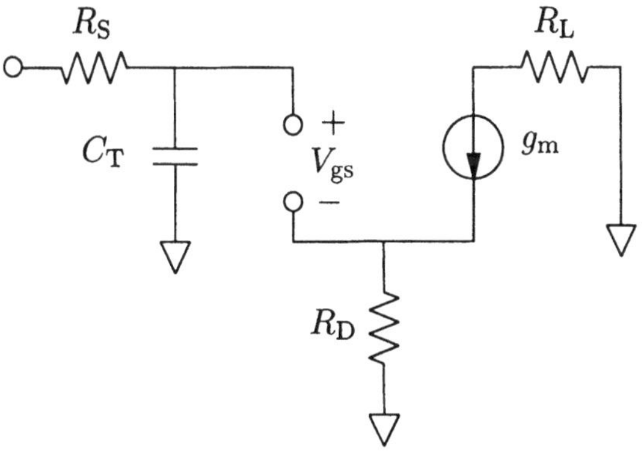

Figure 2.15. Modeling dynamics using the Miller approximation.

ure 2.15) is the recognition that the dynamics of a single stage are dominated by a single pole, which arises from the interaction between source resistance R_S and the input capacitance. With no source degeneration, this input capacitance would be the sum of C_{gs} and the Miller multiplied C_{gd}:

$$C_T = C_{gs} + (1 + g_m R_L)C_{gd} \simeq C_{gs} + (g_m R_L)C_{gd}, \qquad (2.57)$$

where have made the approximation that the gain, $A_v = g_m R_L$, is significantly greater than unity. This capacitance, together with the source resistance, creates a pole with time constant τ:

$$\tau = R_{\mathrm{S}}(C_{\mathrm{gs}} + (g_{\mathrm{m}}R_{\mathrm{L}})C_{\mathrm{gd}}). \tag{2.58}$$

Source degeneration causes the real part of the impedance looking into the gate to increase. At frequencies below the transistor's f_{T}, however, the capacitive part still dominates and we replace C_{gs} in the Miller formulation with

$$\frac{C_{\mathrm{gs}}}{1 + g_{\mathrm{m}}R_{\mathrm{D}}}. \tag{2.59}$$

Source degeneration reduces the gain of the stage from $g_{\mathrm{m}}R_{\mathrm{L}}$ to

$$\tilde{A}_{\mathrm{v}} = \frac{g_{\mathrm{m}}R_{\mathrm{L}}}{1 + g_{\mathrm{m}}R_{\mathrm{D}}}. \tag{2.60}$$

Our C_{T} capacitance is accordingly modified to

$$\tilde{C_{\mathrm{T}}} \simeq \frac{C_{\mathrm{gs}}}{1 + g_{\mathrm{m}}R_{\mathrm{D}}} + \left(\frac{g_{\mathrm{m}}R_{\mathrm{L}}}{1 + g_{\mathrm{m}}R_{\mathrm{D}}}\right)C_{\mathrm{gd}}. \tag{2.61}$$

(We continue to assume that $\tilde{A}_{\mathrm{v}}$ is much greater than unity.) Finally, it can be seen that the effect of feedback has been to reduce the time constant of the pole by a factor of the return difference $1 + g_{\mathrm{m}}R_{\mathrm{D}}$, exactly as was shown in section 2.2.4:

$$\tilde{\tau} = R_{\mathrm{S}}\tilde{C_{\mathrm{T}}} = \frac{\tau}{1 + g_{\mathrm{m}}R_{\mathrm{D}}}. \tag{2.62}$$

If we define τ as in equation 2.58, it can be seen that the source-coupled pair maps perfectly to figure 2.4.

Finally, note that the dynamics here, which are the poles formed by the output impedance of stage i with the input capacitance of stage $i+1$, *are* the interstage loading effects.

2.7.3.1 An alternative formulation: open-circuit time constants

For bandwidth optimization in pure circuit systems, it is often useful to use the method of open-circuit time constants. The method may be summarized as computing the resistance R_i seen across the terminals of capacitors C_i with all other capacitors considered open circuits. The frequency $\omega = \frac{1}{\sum_i R_i C_i}$ has been shown to be a good estimate of the -3dB bandwidth for many common circuits. Moreover this estimate, when applicable, is usually conservative. We direct the interested reader to the excellent discussions in [8] and [10].

We present this method as an alternative. For a given stage, the open-circuit resistance for C_{gd} can be shown to be

$$R_{\mathrm{gd}} = R_{\mathrm{S}} + R_{\mathrm{L}} + \frac{g_{\mathrm{m}}}{1 + g_{\mathrm{m}}R_{\mathrm{D}}}R_{\mathrm{S}}R_{\mathrm{L}}, \tag{2.63}$$

a simple posynomial in the design variable $l = 1 + g_m R_D$. For C_{gs}, the result is

$$R_{gs} = \frac{R_S + R_D}{1 + g_m R_D}, \tag{2.64}$$

which we write as the posynomial in l

$$R_{gs} = \frac{1}{g_m} + \left(R_S - \frac{1}{g_m}\right) l^{-1}. \tag{2.65}$$

2.7.4 Static noise model

Two sources of noise in MOSFETs, drain noise and gate noise, share a common physical origin [10]. Figure 2.16 shows their places in the MOSFET model. Their corresponding places in our theoretical framework are clear, and

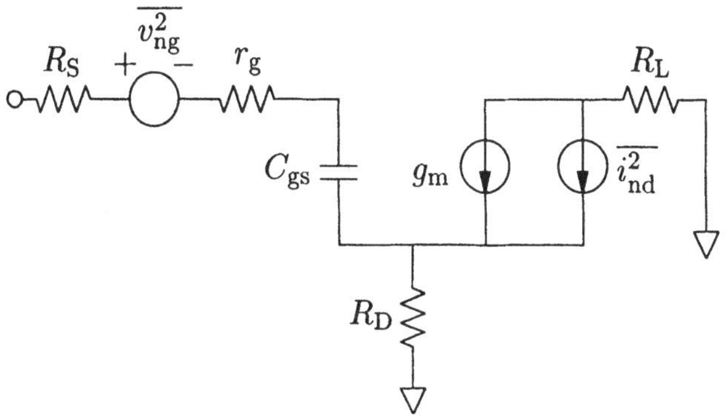

Figure 2.16. MOSFET noise model.

shown in figure 2.17. We only show the noise sources associated with the

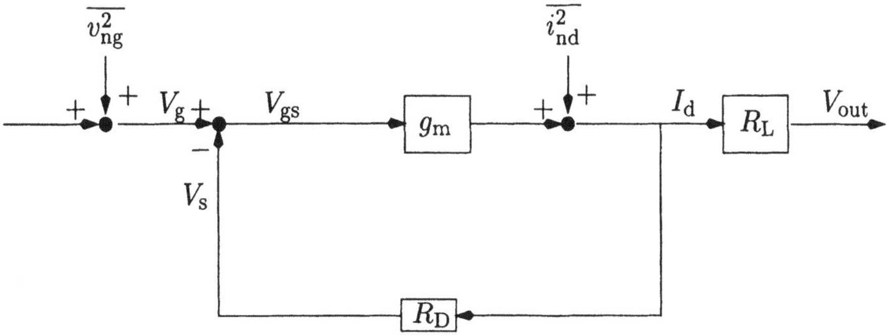

Figure 2.17. MOSFET gate and drain noise.

active device itself. Resistors are known to introduce noise as well, and their contribution is straightforward to include. The noise of R_L, for example, is naturally incorporated as part of the gate noise of the following stage. Similar manipulations can be done for R_D and, of course, R_S.

2.8 Local feedback allocation for power amplifier linearization

This chapter gives the kind of news that is exciting to a design engineer. For the work of manipulating a problem into the appropriate form, one is rewarded with a design process that can have only one of two outcomes: a provably optimum solution, or proof that the problem is infeasible. Such a design process is *extremely rare* in any type of engineering, and is to be sought out wherever possible. It was hoped that the local feedback allocation process could be of use in linearizing RF power amplifiers.

Sadly, local feedback, at least in terms of degeneration, is not a useful technique in RF power amplifiers. The use of feedback presupposes the ready availability of excess gain, and is therefore a resource to be traded for desensitivity to gain variations in the forward path.[3] At frequencies in the low gigahertz range, it is not usually true that excess gain is available. This state of affairs spoils attempts to apply the feedback techniques that are so useful at lower frequencies.

More flaws with degeneration emerge as one examines specific strategies. Resistive degeneration fails because it introduces dissipative power loss, and because it lowers the already low voltage headroom available in modern analog technologies. Inductive degeneration, either alone or as part of a parallel LC tank, doesn't have these problems, but still only linearizes by throwing away precious RF gain. Most of all, though, with degeneration there is no real way to exploit the narrowband nature of RF signals by applying feedback exclusively in the band of interest. The closest one can come is an LC tank as the degeneration element, and the difficulty is that truly narrowband operation (using a high-Q tank) results in more of a notch filter than an amplifier.

Local feedback still has a place in general RF amplifier design. For power amplifiers, however, there are more suitable techniques for achieving linearity and power efficiency. These techniques are the subject of chapter 3.

[3]The way to understand how broadband amplifiers are possible with dominant-pole compensated op-amps, and how nonlinear output stages can be used to effect a linear amplifier, is to take this view of feedback. For op-amps the gain in the forward path is frequency-dependent. Feedback suppresses the variation, and a broadband amplifier is the result. Similarly, nonlinear output stages represent gain variations that are amplitude-dependent, and suppressing these variations is equivalent to "linearizing" the amplifier.

Chapter 3

THE PROBLEM OF LINEARIZATION

The extreme desirability of achieving linearity *and* power efficiency in a power amplifier has motivated a great deal of research into linearization techniques. The basic idea is to run the power amplifier as close to saturation as possible to maximize its power efficiency, and then employ some linearization technique to suppress the distortion introduced in this near–saturated region. This chapter gives a brief overview of the linearization problem, as well as of some of the linearization techniques that are in the current literature.

3.1 The tradeoff between linearity and power efficiency

The impact of the power amplifier on transceiver performance is twofold. First, because it consumes the lion's share of the power budget, the power efficiency of the entire system is improved almost exactly to the extent that such improvement is made with the power amplifier. In a cellular phone, efficiency leads to longer battery lifetimes. In a basestation, efficiency can mean reduced thermal management problems. Second, the power amplifier in large part determines the transceiver's ability to use the spectrum efficiently. If the goal, for instance, is to transmit as many bits per second given a fixed RF channel width, the system designer will want to use a sophisticated modulation technique. The best of these modulation techniques require a highly linear PA.

The pressing question is: do linearity and power efficiency trade off for fundamental reasons? It may be impossible to answer that question definitively. It is always possible, for instance, that there exists a power transmission technique unlike any that we have seen, which achieves linearity and power efficiency with little effort from the designer. Narrowing the question to "circuit realizations based on nonlinear, three-terminal devices," we can at least say that no one has yet found a way to achieve this elusive goal. For reasons that may as well be fundamental, then, linearity and power efficiency are competing design objectives.

Understanding of this state of affairs begins with the observation that a perfect switch dissipates no power. When a switch is open, the current between the terminals is zero regardless of the voltage across them, and when a switch is closed, the voltage across the terminals is zero regardless of the current between them. The result in either case is a zero IV product, and therefore zero power dissipation in the switch.

Figure 3.1 shows a simplified, high-efficiency power amplifier driven to act as a switch. The inductor acts is an RF choke, blocking RF signals while

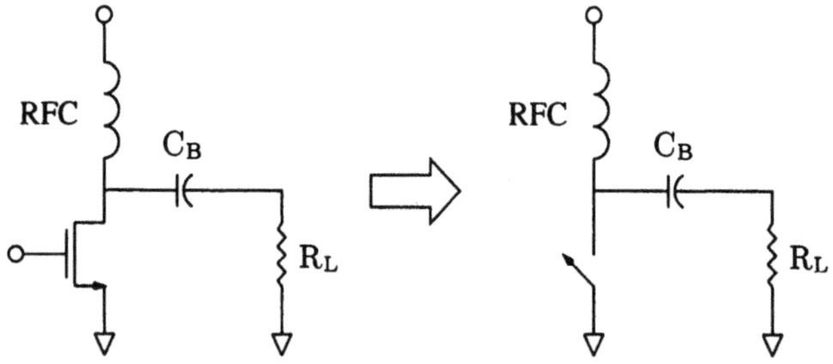

Figure 3.1. A high-efficiency power amplifier.

carrying a DC bias current, and C_B is a large capacitor solely for the purpose of AC coupling. In order to get the transistor to imitate the perfect switch on the right, its gate must be driven very hard. Signals at the gate that slam between the supply rails are not uncommon, and sometimes an even more aggressive drive is needed. If the result is perfect switching action, though, no power is dissipated in the transistor. The inductor and capacitor are purely lossless elements, and we conclude from conservation of energy than any power drawn from the supply must be dissipated in the load. This is the very definition of 100% efficiency. It must be remembered that the key to achieving this high efficiency is driving the transistor hard, with a high voltage signal at the gate.

By contrast, linearity in power amplifiers is most easily accomplished by driving the transistor gently, with as small a signal as possible driving the gate. The mathematical reason for this is discussed in section 3.3.1. What one discovers is that the harder the transistor is driven, the greater its nonlinear transfer characteristic asserts itself. The power efficiency of the circuit approaches zero as the input drive goes to zero, and approaches 50% as the sinusoidal input drives the output into saturation.[1] Since we cannot at once drive a power am-

[1]The efficiencies greater than 50% characteristic of switching power amplifiers derive from, in addition to high amplitude input drive, forcing the output to approximate a square wave.

plifier gently and brutally, we are forced to accept efficiency and linearity as competing design objectives. It is the severity of this tradeoff that we try to soften using linearization techniques.

3.2 Can nonlinear system theory help?

There have been some efforts to apply nonlinear system theory, particularly the theory of Volterra series [17], to power amplifier linearization [18]. Investigations in this vein often crush themselves under the weight of cumbersome, unwieldy analysis and calculation, leading to little or nothing in the way of insight. This is in wild contrast to linear system theory, in which analysis and insight are often tightly coupled. We are fortunate not only that linear system theory exists, but also that nature allows us to approximate so many systems as linear (and time invariant).

A casual study of nonlinear system theory has some relevance to the power amplifier problem, however. From it we find that it is the odd-order terms in the nonlinear transfer polynomial that cause distortion products that lie in-band. For *time-invariant* nonlinear systems, subharmonic generation is not possible. Most of all, though, we gain an appreciation for the nightmare that is power amplifier modeling. A linearization technique that does not require a detailed model of the power amplifier, then, should be regarded as having a fundamental advantage.

3.3 An overview of linearization techniques

While there are many different linearization systems described in the open literature, for the most part they fall into one of eight categories: power backoff, predistortion, adaptive predistortion, feedforward, dynamic biasing, envelope elimination and restoration (EER), linear amplification with nonlinear components (LINC), and Cartesian feedback. There is also polar feedback. However, EER serves as a sufficiently illustrative and common example of this technique that it will not be described separately. In this section, each of these categories is briefly reviewed.

3.3.1 Power backoff

Power backoff exploits the observation that any amplifier appears linear for sufficiently small departures from its bias condition. This is the starting point for analyzing transistor amplifiers in any undergraduate course on the subject, where it bears the name "small-signal analysis." The principle is best illustrated mathematically. Consider that the nonlinear transfer characteristic of any device can be expanded as the Taylor series

$$I_{\text{out}} = I_0 + a_1 v_{\text{in}} + a_2 v_{\text{in}}^2 + \cdots , \tag{3.1}$$

where I_{out} is the total output current, I_0 is the bias output current, and v_{in} is the signal input voltage. It is seen that by shrinking v_{in}, the linear term can be made to dominate all but the DC term I_0. For RF power amplifiers, of course, this DC term is easily blocked, and it remains that if we drive the amplifier gently enough, the system will not suffer the spectral regrowth that accompanies nonlinear amplification. The decision to shrink v_{in} is called "backing off," and an engineer might speak, for instance, of "backing off from the 1dB compression point until we meet our linearity spec."

Power backoff is not really true to the spirit of linearization techniques, because no attempt is made to beat the linearity-vs.-efficiency tradeoff. The input drive is simply reduced until linearity requirements are met, and the resulting efficiency is accepted.[2] It is mentioned here because it is so commonly used, and because it is, in fact, the best solution in some cases.

3.3.2 Predistortion

Given a precisely specified nonlinear characteristic for a power amplifier, $p(x)$, one approach is to pose the linearization problem as follows:

> Choose a predistorter, $p_{\text{pre}}(x)$, and/or a postdistorter, $p_{\text{post}}(x)$, such that the cascade, $p_{\text{post}}(p(p_{\text{pre}}(x)))$, is a linear function.

While this is the most general, most rigorous way to pose the problem in theory, in practice it is rare for a postdistortion solution even to be considered. Adding a postdistorter in the path between the power amplifier and the antenna would almost certainly introduce a disastrous amount of loss.

The corrective distortion concept most often finds expression as the predistorters of figure 3.2. The first implementation (the top half of the figure) shows predistortion applied to the RF signal itself. This predistortion block is typically an analog, clever realization of the nonlinear inverse of the power amplifier transfer characteristic (for examples, see [19] and [20]). The second type of predistortion is applied to the baseband symbols before upconversion, and is illustrated in the lower half of figure 3.2 [21]. In this implementation, the predistorting function is typically maintained as a digital look-up table.

Predistortion's chief attribute is its conceptual simplicity. It also does not suffer a bandwidth limitation that is as stringent as some feedback techniques. Its chief liability that it is helpless in the face of variations in the power amplifier. The exact nonlinear characteristic of the power amplifier varies with temperature and process variations, not to mention aging. As conditions change, a fixed predistorter will tend to become misaligned with the nonlinearity of the power amplifier, resulting in suboptimal performance. Another drawback to predistortion is its dependence on a good model of the power amplifier. Power

[2]The efficiencies can be extremely low. It is not uncommon to hear figures on the order of 5%.

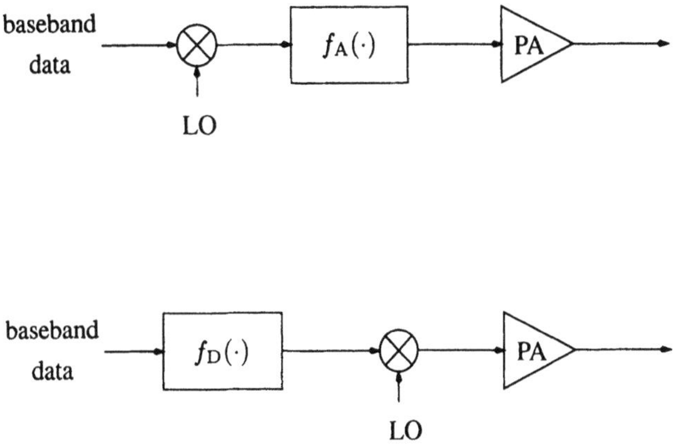

Figure 3.2. Using predistortion to linearize a power amplifier.

amplifiers are notoriously difficult to model, and predistortion immerses the designer directly into this thorny issue.

3.3.3 Adaptive predistortion

Figure 3.3 is an example of adaptive predistortion, a logical evolutionary step forward from ordinary predistortion. The idea is to solve predistortion's

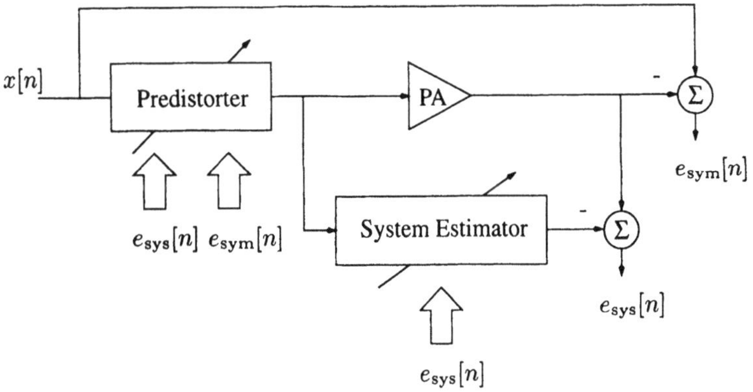

Figure 3.3. An example of adaptive predistortion.

greatest weakness, which is its inability to cope with variations in the power amplifier. The system shown in figure 3.3[22] maintains a dynamically updated model of the power amplifier. Based on the parameters of this model, and on comparisons between the transmitted data and the PA output, a predistortion table is computed. By periodically updating the predistortion table, the system is able to react to changes due to process variations, temperature, and the like. A substantial improvement comes without incurring a bandwidth limitation.

Complexity is adaptive predistortion's main drawback. This strategy is usually implemented with considerable computing power, often in the form of a digital signal processor. The power required for the DSP can effectively rule out adaptive predistortion as an option for portable transceivers.

Adaptive predistortion shares with predistortion the difficulty of needing to carefully model the power amplifier. In the case of the former, the difficulty is compounded by the discrete-time stability problem associated with convergence of the model parameters according to the adaptive algorithm.

3.3.4 Feedforward

Figure 3.4 illustrates feedforward linearization [23]. Conceptually, the delay

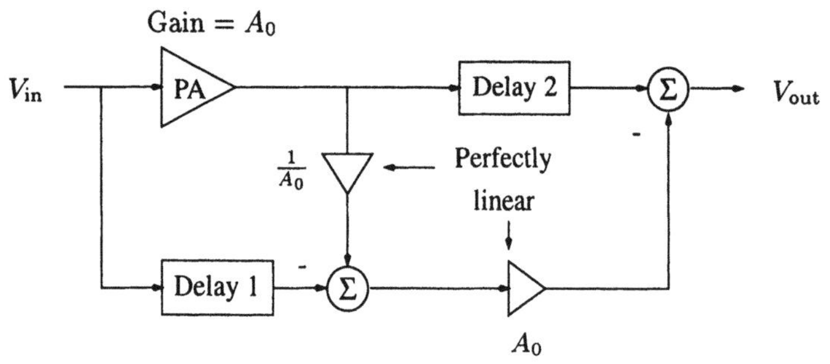

Figure 3.4. Feedforward linearization.

blocks "delay 1" and "delay 2" can be ignored at first. It is seen that the input to the perfectly linear amplifier of gain A_0 is intended to represent the additive distortion products present at the output of the PA. These products, after being scaled back up to their original magnitudes, are then subtracted from the PA output. The theoretical result is a linear power amplifier, and there is no bandwidth limitation that is directly attributable to the linearization effort.

As always, things are not so neat in practice. The same lack of feedback that gives it superior bandwidth performance gives it the potential to be a very high-maintenance system. Because of phase shifts introduced in the carrier by the PA and other amplifiers, for instance, delays 1 and 2 must be very tightly tuned in order to realize good performance. The gains of the linear amplifiers also must be very well matched to the gain of the PA. Both the phase and scaling adjustments are bound to be disrupted by any sort of drift or process variations, unless they can somehow be made to track each other. Finally, the analog delay and subtractor in the output path of the power amplifier must be extremely low-loss in order for the whole linearization effort to be worthwhile.

3.3.5 Dynamic biasing

It sometimes happens that an amplifier exhibits good linearity over a wide range of input signals. When this is true, the focus of the linearization effort shifts away from making great gains in linearity. Instead, one thinks of trading the excess linearity for a boost in power efficiency. A strategy for doing this is called dynamic biasing [24].

In a dynamic biasing scheme, the gate (or base) bias voltage of the amplifying device is varied according to the known statistics of the transmitted signal. The goal, at all times, is for the amplifier to use the minimum bias current needed for a given level of linearity. While it is difficult to achieve this ideal in practice, substantial gains in efficiency are nevertheless possible.

3.3.6 Envelope elimination and restoration

The unmatched efficiency of switching power amplifiers has motivated efforts to use them for linear amplification. Envelope elimination and restoration (EER) is one such effort, and is illustrated in figure 3.5[25]. The basic idea is to

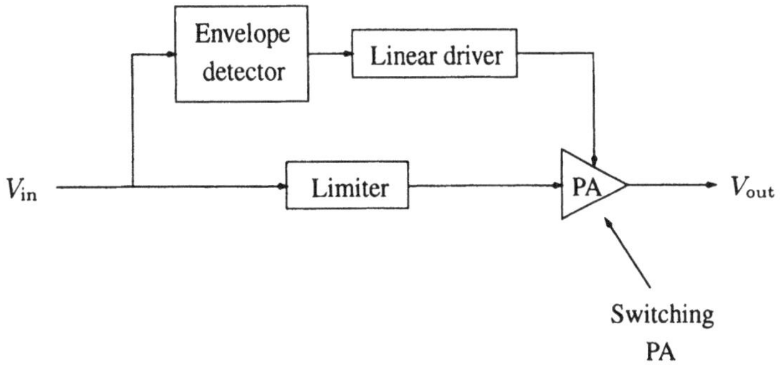

Figure 3.5. Envelope elimination and restoration.

impose the varying envelope on the power supply of the switching PA, typically using a switching, DC-DC feedback power converter. The reader will notice in this example that no feedback is employed to ensure the fidelity of the phase part of the carrier. This is not always the case. Sometimes the phase is regulated by a phase-locked loop, and when the envelope and phase are feedback-regulated the system is said to employ *polar* feedback[18].

The DC-DC converter represents a design issue that can make EER and polar feedback systems difficult to implement. An excellent efficiency figure for a DC-DC converter is 80%. If the power amplifier achieves a drain efficiency of 60% (also impressive), the overall energy efficiency drops to 48%. High efficiency in the power converter is difficult to maintain for high bandwidth signals, a problem exacerbated by the fact that the polar representation has a higher bandwidth than the Cartesian component representation[26]. The net result is that the efficiency burden of the power amplifier has been shifted back to the power converter, and it requires careful design to produce an overall benefit.

A general concern unique to strategies that deal in polar symbols is synchronization between the amplitude and phase signals as they are imposed on the RF carrier. At the time of this writing, there is no way to enforce this synchronization in a closed-loop fashion. The current state of the art is to minimize the worst-case relative delays between the envelope and phase signal paths.

3.3.7 LINC

Another approach to harnessing the efficiency benefits of switching PA's is illustrated in figure 3.6[27, 28, 29]. Referred to as "linear amplification

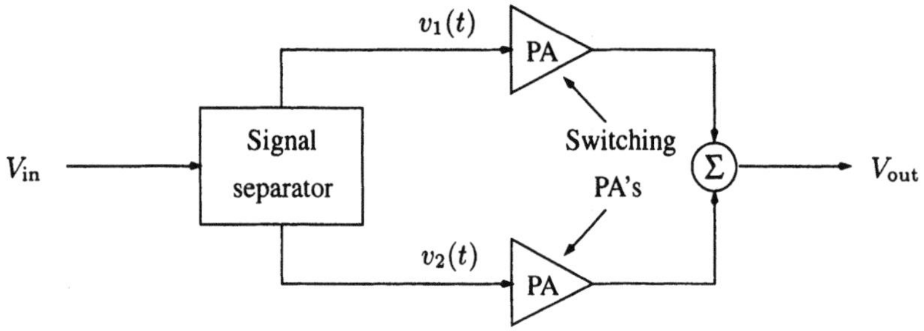

Figure 3.6. The LINC concept.

with nonlinear components," or LINC, the idea is that if properly chosen, two constant-envelope signals driving the separate power amplifiers can, when summed, result in the general bandpass signal $v_{in}(t) = a(t)cos[\omega_c t + \phi(t)]$. The proper choices for these constant-envelope signals turn out to be $v_1(t) = 0.5V_0 \sin[\omega_c t + \phi(t) + \theta(t)]$ and $v_2(t) = -0.5V_0 \sin[\omega_c t + \phi(t) - \theta(t)]$, where $\theta(t) = \sin^{-1}[\frac{a(t)}{V_0}]$.

Conceptually, LINC is such a compelling idea that the pressing question arises: why isn't it the only solution in use today? The most obvious objection might be the complexity involved in generating $v_1(t)$ and $v_2(t)$, but this can be accomplished with sufficient computing power. A very real difficulty is managing the phase and gain mismatch between the two signal paths, and the drift thereof. The most prohibitive barrier, though, is implementing the summation in a way that is low loss *and* maintains high isolation between the two PA's. At the time of this writing, this last problem remains unsolved.

3.3.8 Cartesian feedback

Much of the work described in this book focuses on this last technique, Cartesian feedback[23, 18, 30, 32, 33, 34, 31, 35, 36, 37, 38]. It is called Cartesian feedback because the feedback is based on the Cartesian coordinates of the baseband symbol, I and Q, as opposed to the polar coordinates. The concept is illustrated in figure 3.7.

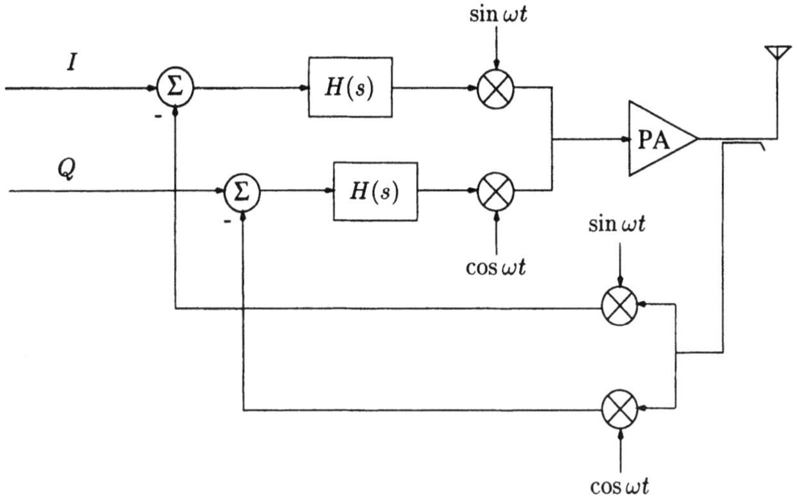

Figure 3.7. Cartesian feedback.

Fundamentally, the concept behind this system is negative feedback. A couple of factors complicate its expression in the context of an RF transmitter, however. The first is the extremely high frequency of many RF carriers, with modern standards calling for frequencies on the order of a few gigahertz. At this time, it is virtually impossible to build a high-gain, stable analog feedback loop with a crossover frequency in that range. The second factor is the recognition that in modulating an RF carrier, we are not shaping a voltage waveform in its entirety. Instead, we are shaping two independent characteristics of that carrier.

Cartesian feedback's way of dealing with the first factor is the inclusion of a frequency translation step in the feedback path, shown as a downconversion mixer in figure 3.7. The loop is then closed at baseband, rather than at the carrier frequency. The system consequence is to linearize only in a narrow band of the spectrum centered about the carrier, rather than from DC to the carrier. This is an ingenious way to exploit the narrowband nature of most RF signals.

The second factor manifests as the "double loop" structure of the system. There are two degrees of freedom in shaping, or modulating, an otherwise free-running RF carrier, and at least two choices of coordinate systems that fully describe the modulation. For polar feedback and EER, the choice made is to consider an RF carrier as having an amplitude and a phase. The structure of a polar feedback system reflects this choice, having one control loop for the amplitude, and another for the phase. An equivalent choice of coordinates is the Cartesian components, in which we consider the modulated carrier as the sum

$$A(t)\sin(\omega_0 t + \phi(t)) = I(t)\sin\omega_0 t + Q(t)\cos\omega_0 t,$$

where

$$I(t) = A(t)\cos\phi(t)$$

and

$$Q(t) = A(t)\sin\phi(t).$$

It is seen that Cartesian feedback treats the two degrees of freedom in a symmetrical way, allowing the structure of the system to take the form of two identical loops. This is in direct contrast to polar feedback, where the two degrees of freedom must be treated very differently.

A full discussion of the strengths and drawbacks of Cartesian feedback is left for section 5.1.

Chapter 4

PHASE ALIGNMENT IN CARTESIAN FEEDBACK SYSTEMS

The importance of phase alignment in Cartesian feedback systems is emphasized in nearly all of the writing on the subject. No compact solution to the problem of maintaining phase alignment has emerged in the literature, however, and this absence of a solution has prevented the use of Cartesian feedback. And while it is generally known that stability and phase alignment are related, a clear mathematical analysis of that relationship is also conspicuously absent from the literature.

The purpose of this chapter is to begin to fill these two gaps in the literature. A new mathematical treatment of the consequences of phase misalignment is given first, followed by a nonlinear control method for maintaining phase alignment. A discrete-component prototype based on these ideas, described in detail in appendix A, demonstrates the soundness of the theory while identifying the circuit nonidealities to which it is sensitive. Finally, as a way of overcoming the shortcomings of that first prototype, a new technique for doing analog multiplication is proposed, which is employed in the improved phase alignment system of chapter 5.

4.1 Consequences of phase misalignment in Cartesian feedback systems

Figure 4.1 shows a typical Cartesian feedback system [23]. The system block $H(s)$ represents the loop driver amplifiers, which provide the loop gain as well as the dynamics introduced by the compensation strategy. The loop drivers feed the baseband inputs of the upconversion mixer, which in turn drives the power amplifier. Some means of coupling the output of the power amplifier to the downconversion mixer is employed, and the output of this mixer is used to close the feedback system.

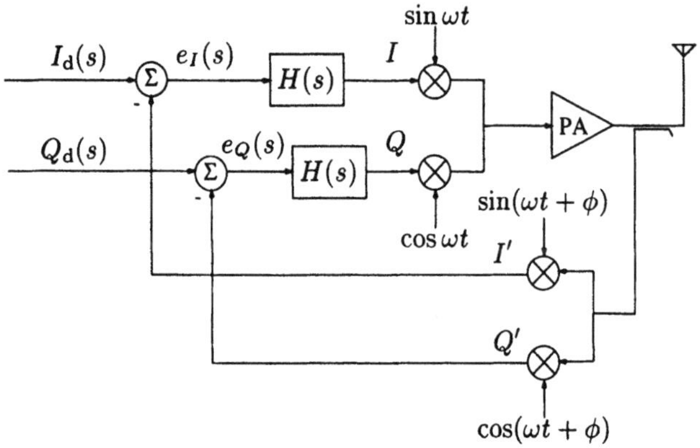

Figure 4.1. Typical Cartesian feedback system.

4.1.1 Terminology Convention

We first identify the terminology conventions that are used in the analysis that follows. Figure 4.2 shows an example feedback system. The signal $e(s)$

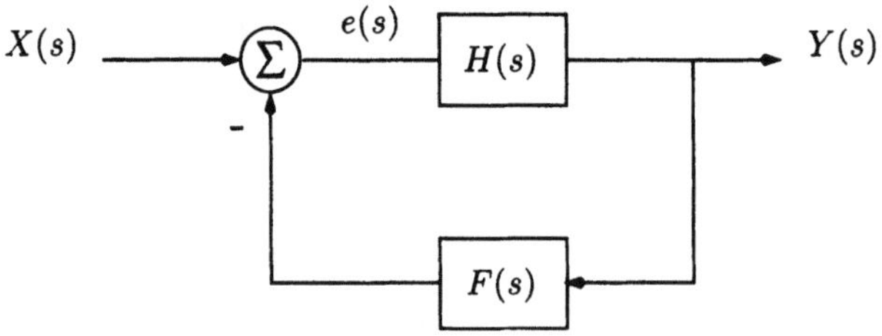

Figure 4.2. Simple feedback system.

is the error, or difference between the command input, $X(s)$, and the feedback signal. The output of this system, $Y(s)$, is related to the command input through the well-known relation

$$\frac{Y(s)}{X(s)} = \frac{H(s)}{1 + H(s)F(s)}.$$

We refer to the quantity $H(s)F(s)$ as the loop gain, or loop transmission, of the system, and give it the symbol $L(s)$.

4.1.2 Impact of phase misalignment on stability

Ideally, a Cartesian feedback system functions as two identical, decoupled feedback loops: one for the I component, and one for the Q component. This corresponds to the case of $\phi = 0$ in figure 4.1. In practice, however, this state of affairs must be actively enforced. Delay through the power amplifier, phase shifts of the RF carrier due to the reactive load of the antenna, and mismatched interconnect lengths between the local oscillator (LO) source and the two mixers all manifest as an effective nonzero ϕ. Worse, the exact value of ϕ varies with temperature, process variations, output power, and carrier frequency. A Cartesian feedback system in which ϕ is nonzero is said to have phase misalignment. In this state the two feedback loops are coupled, and the stability of the system is compromised.

The impact of phase misalignment on system stability can be seen mathematically. We start by observing that the demodulated symbol S' is rotated relative to S by an amount equal to the phase misalignment ϕ. To see this, we write Cartesian components of the demodulated symbol

$$\begin{aligned} I' &= (I \sin \omega t + Q \cos \omega t) \sin(\omega t + \phi) \\ Q' &= (I \sin \omega t + Q \cos \omega t) \cos(\omega t + \phi), \end{aligned}$$

where ω is the carrier frequency. Using trigonometric identities and assuming frequency components at 2ω are filtered out, we arrive at S'

$$I' = \frac{1}{2}(I \cos \phi + Q \sin \phi) \tag{4.1}$$

$$Q' = \frac{1}{2}(-I \sin \phi + Q \cos \phi). \tag{4.2}$$

We see that for $\phi \neq 0$, an excitation on the I input of the modulator results in a signal on the Q' downconverter output (and similarly for Q and I'). Accordingly, we say that the two loops are coupled.[1]

One method of stability analysis is to consider the error signals $e_I(s)$ and $e_Q(s)$ shown in Figure 4.1. Recall that for a single feedback loop, the error signal is written

$$e(s) = \frac{X(s)}{1 + L(s)},$$

where $X(s)$ is the command input. In the present case, let the phase misalignment be ϕ. Furthermore, we set $Q_{\mathrm{d}} = 0$ without loss of generality.[2] The error expressions, as a function of the single input $I_{\mathrm{d}}(s)$, are written

$$\begin{aligned} e_I(s) &= I_{\mathrm{d}}(s) - L(s)e_I(s) \cos \phi - L(s)e_Q(s) \sin \phi \\ e_Q(s) &= L(s)e_I(s) \sin \phi - e_Q(s)L(s) \cos \phi, \end{aligned}$$

[1]Technically, $\phi = \pi$ is also an uncoupled case. However, there is now an inversion in both loops, resulting in positive feedback instead of the desired negative feedback.

[2]We do not lose generality as long as we stay with linear analysis.

where $L(s)$ includes the dynamics of the loop compensation scheme $H(s)$ and the (linearized) dynamics introduced by the modulator, power amplifier, and demodulator. From here, it is straightforward to show that

$$e_I(s) = \frac{X(s)}{1 + L(s)\cos\phi + \frac{[L(s)\sin\phi]^2}{1+L(s)\cos\phi}}.$$

This reduction of the system to a single-input problem now yields considerable insight. We identify an effective loop transmission, $L_{\text{eff}}(s, \phi)$, as follows:

$$L_{\text{eff}}(s,\phi) = L(s)\cos\phi + \frac{[L(s)\sin\phi]^2}{1 + L(s)\cos\phi}. \tag{4.3}$$

For perfect alignment, $\phi = 0$ and L_{eff} is simply $L(s)$. The worst alignment is $\phi = \frac{\pi}{2}$, for which $L_{\text{eff}} = [L(s)]^2$ and so the loop dynamics are a cascade of the dynamics in the uncoupled case. Unless designed with this possibility in mind, most choices of $H(s)$ yield unstable behavior in this second case. Equation 4.3 shows that traditional measures of stability degrade continuously as ϕ sweeps from 0 to $\frac{\pi}{2}$, a fact demonstrated experimentally by Briffa and Faulkner [34].

4.1.3 Compensating the system for robustness to phase misalignment

Equation 4.3 offers a great deal of insight into what happens in a phase-misaligned Cartesian feedback system. Physically, the fully coupled ($\phi = \frac{\pi}{2}$) case behaves as depicted in figure 4.3, where $P(s)$ represents the dynamics that the upconversion mixer, power amplifier, and downconversion mixer contribute to the loop transmission. In the literature, all efforts with regards to the phase

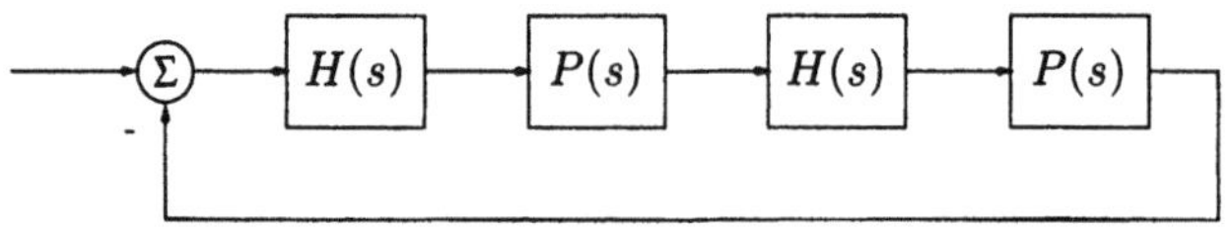

Figure 4.3. Cartesian feedback under 90-degree misalignment.

alignment problem have focused, naturally, on ensuring phase alignment. But there is at least one other approach that deserves consideration: is it possible to choose $H(s)$ such that it is stable for large phase misalignments?

The answer depends in part on what one means by "large." Considering a misalignment of $\phi = \pi$, for instance, is discouraging. In this case $L_{\text{eff}} = -L(s)$, and there is simply no compensation strategy that is indifferent to the sign of the loop transmission. Cartesian feedback in fact does become a positive feedback system for misalignments in the open interval $(\frac{\pi}{2}, \frac{3\pi}{2})$, where the exact point of transition from negative to positive feedback depends on the details of $L(s)$. To avoid considering positive feedback cases, then, it is sensible to restrict the range of misalignments to the closed interval $[-\frac{\pi}{2}, \frac{\pi}{2}]$.

That stability margins degrade *continuously* with ϕ suggests that finding a compensation strategy that works in the limiting cases of $\phi = 0$ and $\phi = \frac{\pi}{2}$ will solve the problem for the whole interval. Assuming the dynamics of the loop are dominated by $H(s)$, a compensation strategy that emerges is

$$H(s) = \frac{k}{s^x},$$

where $0 < x < 1$. Such "slow-rolloff" functions, while not truly realizable with a lumped-element network, can be approximated by alternating poles and zeros such that the average slope of H(s) is the appropriate dB-per-decade[3]. In the case of $x = 0.5$, for instance, stability as measured by phase margin would be excellent: 135 degrees in the aligned case, and 90 degrees in the $\frac{\pi}{2}$ misaligned case.

Root locus analysis confirms that slow-rolloff compensation is a viable approach to designing for large misalignments. Figure 4.4 shows the root loci for

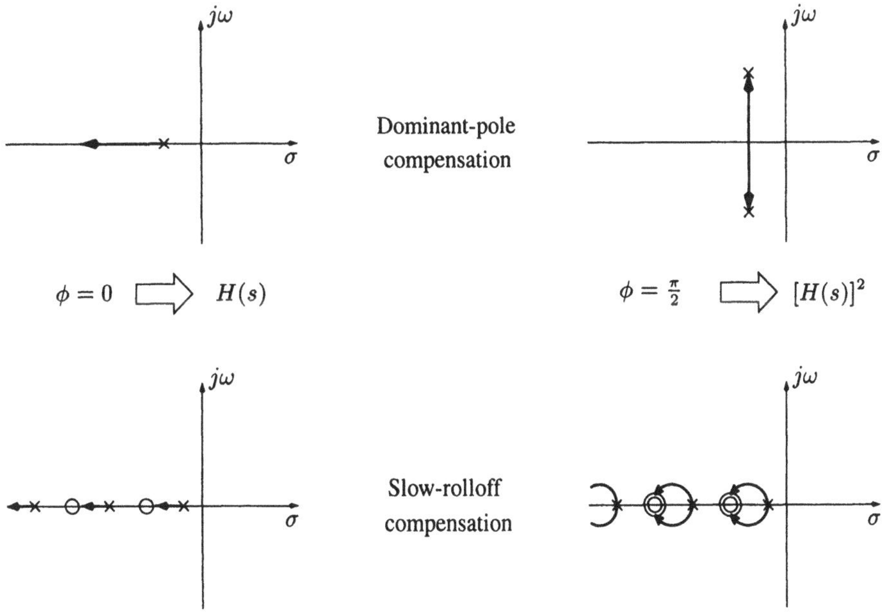

Figure 4.4. Root locus plots for dominant-pole and slow-rolloff ($\frac{k}{\sqrt{s}}$) compensation.

the dominant-pole and $\frac{k}{\sqrt{s}}$ compensation strategies. It can be seen that even in the case of the dominant-pole, 90-degree misalignment doesn't necessarily lead to right-half-plane poles. At best, however, the result is a lightly damped, complex pair of poles. At worst, high-frequency poles not shown here (or not modeled) push this complex pair into the right-half plane. By contrast, the slow-rolloff compensation is seen to lead to heavily-damped complex pole pairs, and one expects a corresponding reduction in overshoot and ringing in response to an input step. One also expects the low-frequency, zero-pole doublets of the root loci to manifest themselves as slow-settling "tails" in the step response [3].

Experiments carried out on the final IC in accordance with this compensation discussion validate these expectations. As seen in section 5.5.2.2, the slow-rolloff technique stabilizes the system for all misalignments up to 90 degrees. In addition to shedding light on compensation strategies for Cartesian feedback systems, the importance of these experiments is that they confirm the understanding developed in section 4.1.2.

4.2 A nonlinear regulator for maintaining phase alignment

Occasionally continuous regulation of the phase alignment is not needed, and it suffices to introduce a manually adjustable delay between the LO source, and, say, the demodulator [18]. This approach is only feasible, however, if the system is not subject to variations in temperature, carrier frequency, process parameters, or, in some cases, output power. For cases in which the alignment must be regulated, various methods have been proposed in the literature [30, 32, 36, 37].

We present our control concept as a compact, truly continuous solution to the problem of LO phase alignment. It is truly continuous because it does not, for example, rely on the appearance of a specific symbol or pattern in the outgoing data stream. It is compact because it is easily implemented without digital signal processing, as presented here. This is a particularly compelling advantage, as the signals in a Cartesian feedback system are necessarily in analog form. And we emphasize that, because the concept is based on the processing of baseband symbols, its realization is independent of carrier frequency.

4.2.1 Nonlinear dynamical system

Figure 4.5 represents a baseband symbol at the inputs of the modulator and at the outputs of the demodulator of a Cartesian feedback system. Mathematically the vectors are described in both Cartesian and polar coordinates, with primed coordinates denoting the demodulated power amplifier output and unprimed coordinates denoting the modulator input. In addition to undergoing a distortion in magnitude, the demodulated symbol is rotated by an amount exactly equal to the phase misalignment (see equations 4.1 and 4.2).

A start to the design of a phase alignment regulator is to observe that the signals I, Q, I', and Q', taken together, represent enough information to determine the phase misalignment. Further, they are easily accessible within the

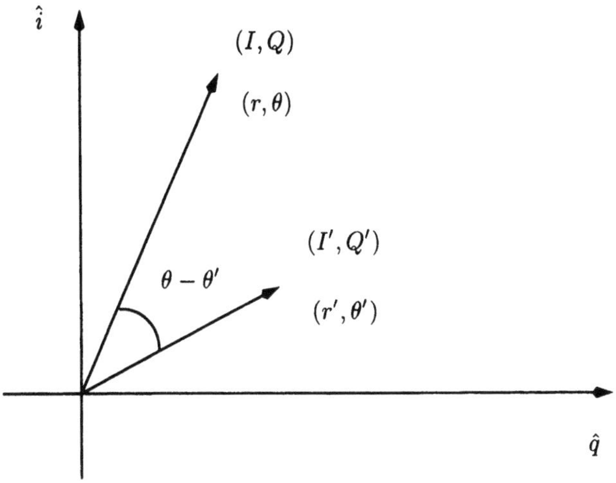

Figure 4.5. Rotation of the baseband symbol due to phase misalignment.

system. We seek to combine these variables such that, over a suitable range, the derived signal is monotonic in the phase misalignment.

One such combining of the variables is the sum of products $IQ' - QI'$. Recognizing that $I = r\sin\theta$ and $Q = r\cos\theta$, and using trigonometric identities, we write the key relation

$$IQ' - QI' = rr'\sin(\theta - \theta'). \tag{4.4}$$

We see that using two multipliers and a subtractor, operations easily realizable in circuit form, one can derive a control signal that is indeed monotonic in the phase misalignment over the range $-\frac{\pi}{2} < \theta - \theta' < \frac{\pi}{2}$.

Figure 4.6 details a nonlinear dynamical controller built around equation 4.4. Using the notation $\Delta\theta = \theta - \theta'$, an implementation can be understood as mechanizing the equation

$$\frac{d\theta}{dt} = -\kappa[r(t)]^2 G\sin(\Delta\theta), \tag{4.5}$$

where κ is a constant of proportionality and gain G is associated with the integrator.

Equation 4.5 presupposes the ability to correct the phase shift by changing θ. The original prototype described in appendix A realizes the required rotation by directly phase shifting the modulator LO. However, substantial power savings result from doing symbol rotation at baseband as shown in figure 4.6. Regardless, rotation should be performed in the forward path of the Cartesian feedback system, where the unavoidable artifacts of imperfect rotation are suppressed.

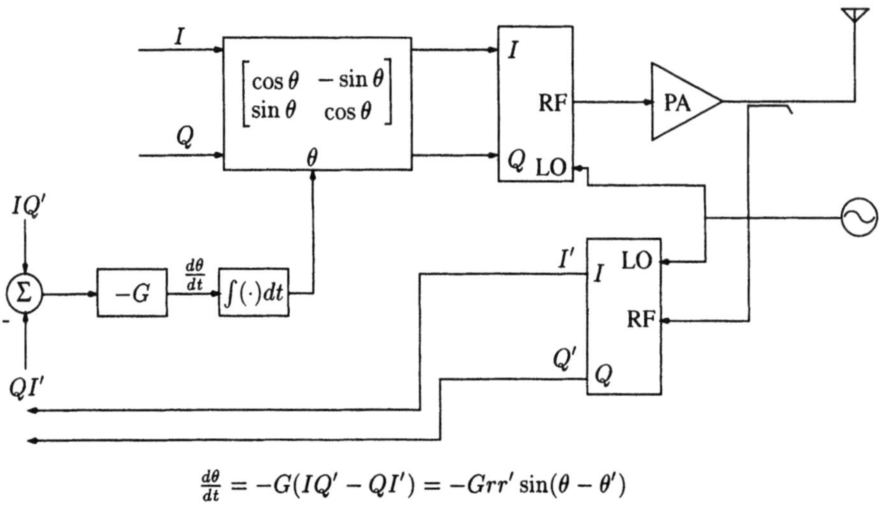

Figure 4.6. Phase alignment concept.

4.2.2 Stability concerns

Our control solution for the phase alignment problem is the simplest of nonlinear dynamical systems. It is seen from equation 4.5 to have two equilibrium points. The first, for which the symbols are aligned, is *stable*. The second, for which the symbols are misaligned by π radians, is unstable. For the ideal system represented by equation 4.5, this is the extent of a rigorous stability analysis.

The real-world situation can be complicated by dynamics associated with the phase shifter (and, possibly, the subtractor). If we provisionally consider a modulation scheme in which the magnitude of transmitted symbols is held constant,[3] $r(t)$ in equation 4.5 loses its time dependence. Linearizing for small phase misalignments, and including the dynamics of the phase shifter as $P(s)$, we can represent the system as shown in figure 4.7. Drawing the system this way requires some manipulation. The output of the phase shifter is not really θ, but rather an additive *part* of θ that gets combined with the polar angle of the symbol being transmitted. However, in the absence of phase distortion and drift, the symbol-by-symbol changes of the polar angle θ are tracked by identical changes in θ'. These symbol-rate changes are thus invisible to an alignment system, and it is appropriate to label the output of $P(s)$ as θ. We

[3]Unlikely when using Cartesian feedback, of course. Temporarily making this assumption, however, yields insight that is broadly relevant to the stability analysis.

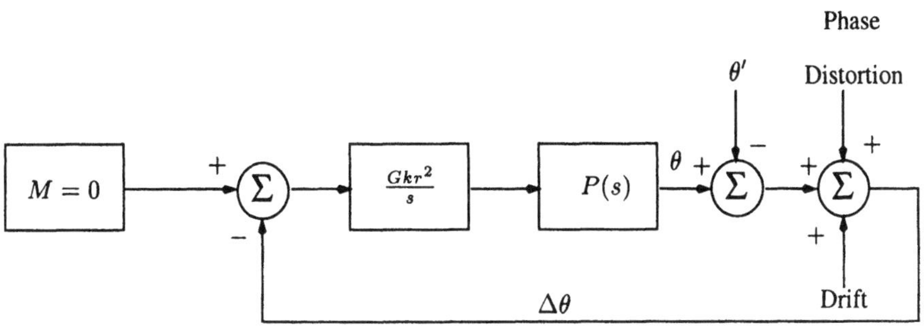

Figure 4.7. Linearized phase regulation system. 'M' is the desired misalignment, which is nominally zero.

can then include the effects of phase distortion and phase alignment drift as the additive disturbances of figure 4.7.

One can ensure stability by choosing G such that, for the largest symbol magnitude, loop crossover occurs before non-dominant poles become an issue. Fortunately, the drift disturbance will normally occur on time scales associated with temperature drift and aging [23]. Suppression of the phase distortion is the domain of the Cartesian feedback itself. It follows that for many systems, little of the design effort need be focused on fast phase alignment.

4.2.3 Quadrature error in the mixers

The analysis of the phase alignment control problem becomes complicated when one considers quadrature error in the mixers. Mathematically, the change is that the Cartesian basis vectors $\hat{\imath}$ and $\hat{q}$ are no longer orthogonal for the mixer with the quadrature error. It can be shown that with this nonideality, no single setting of the phase shifter keeps the quantity $IQ' - QI'$ zero for all possible symbols. This analytical result is independent of the phase alignment method used.

Fortunately, mixers with small quadrature errors ($\pm$ 5 degrees) are easily realized. Such mixers cause no serious problems in the experiments for this work, or in the Cartesian feedback systems that have been described in the literature.

4.2.4 Impact of multiplier offsets

The ability to accurately regulate the phase alignment depends on the ability to accurately calculate $IQ' - QI'$. In this regard, DC offsets associated with the output buffers of the multipliers and the input of the integrator are particularly troublesome. Consider an input-referred offset of δ for the controller integrator, and its effect on the final alignment. We write

$$\frac{d\theta}{dt} = G[-\kappa[r(t)]^2 \sin(\Delta\theta) + \delta] = 0.$$

For the prototype of appendix A, κ is approximately 1.3V^{-1}. For a symbol magnitude of 50mV, we can solve for the offset that results in a 5-degree final misalignment:

$$\delta = (1.3\text{V}^{-1})(50\text{mV})^2 \sin\left(\frac{2\pi(5)}{360}\right) = 283\mu\text{V}.$$

This example illustrates one of the major challenges that the analog multipliers introduce, which is that offsets become increasingly intrusive as symbol magnitudes decrease. A 5-degree misalignment for a symbol magnitude of a volt leads to a δ of 113mV, for example, while for a symbol magnitude of 1mV, δ shrinks to 0.113μV.

Mitigating this effect is the fact the controller slows for smaller signals: until offsets dominate, $|\frac{d\theta}{dt}|$ scales linearly with $[r(t)]^2$. These numbers nevertheless suggest that offsets deserve careful consideration in any circuit realization of this control concept.

4.3 A new technique for offset-free analog multiplication

Section 4.2.4 demonstrates the problems that offsets associated with analog multipliers introduce. We now describe a new technique for analog multiplication , invented specifically for the phase alignment problem, but generally applicable to the problem of analog multiplication. It is first applied in the IC prototype of chapter 5.

An example of a basic, CMOS multiplier cell can be seen in figure 5.9 [39]. A mathematically complete description of a multiplier's offset behavior requires at least three quantities: δ_{I} and δ_{Q}, the offsets attributable to the inputs, and δ_{O}, the offset introduced in the current-to-voltage conversion performed at the output of the multiplier. Minimizing these offsets is difficult. One might imagine, for example, some combination of careful, symmetrical layout and a calibration step. We introduce instead the technique shown in figure 4.8. The underlying idea is to employ chopper stabilization to eliminate (or at least greatly suppress) these multiplier offsets which produce the dominant phase alignment errors in conventional realizations. Long successfully used in precision DC amplifiers [3, 40, 41], two critical modifications are required to apply chopper stabilization to analog multiplication. The first modification is to chop the two inputs in quadrature, and the second is to chop down at *twice* the original chopping frequency. To the extent that this chopping strategy is perfectly implemented, offsets δ_{I}, δ_{Q}, and δ_{O} are completely circumvented.

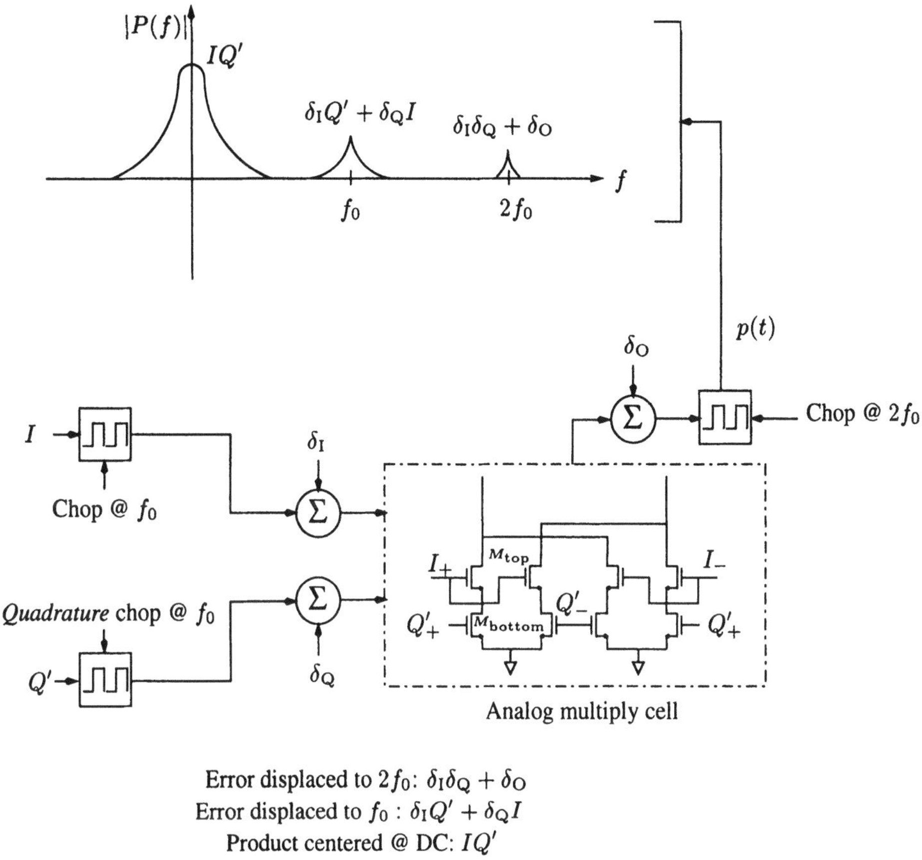

Figure 4.8. New technique for offset-free analog multiplication.

The chopping operation is equivalent to mixing a signal with a square wave of unit amplitude.[4] In the following treatment, one chopping waveform will be denoted $c_0(t)$, and the other, quadrature waveform will be denoted $c_{90}(t)$. We write these two waveforms as their Fourier series decompositions

$$c_0(t) = \sum_{n=0}^{\infty} \frac{4}{(2n+1)\pi} \sin(2n+1)\omega_0 t$$

$$c_{90}(t) = \sum_{m=0}^{\infty} \frac{4}{(2m+1)\pi} (-1)^m \cos(2m+1)\omega_0 t,$$

[4]That is to say, a square wave that alternates between +1 and -1. We clarify because sometimes, particularly in single-ended systems, it is convenient to chop with a square wave that alternates between +1 and 0. This latter case requires a mathematical treatment that differs slightly from what we present here.

where ω_0 is the angular chopping frequency $2\pi f_0$.

Figure 4.8 shows the signals applied to the inputs of the analog multiplier as

$$Ic_0(t) + \delta_I$$

and

$$Q'c_{90}(t) + \delta_Q.$$

Assuming linear multiplication, the output of the multiplier (before the down chopping operation) is written

$$IQ'c_0(t)c_{90}(t) + \delta_I Q'c_{90}(t) + \delta_Q Ic_0(t) + \delta_I\delta_Q + \delta_O. \tag{4.6}$$

The second and third terms of this expression can only have spectral content centered at ω_0 and/or its odd harmonics, while the fourth and fifth terms are centered at DC. The key, then, is to demonstrate that the product $c_0(t)c_{90}(t)$ has spectral content only at even harmonics of the fundamental. A graphical analysis, as shown in figure 4.9, is by far the easiest way to accomplish this. The product $c_0(t)c_{90}(t)$ is seen to be

$$c_0(t)c_{90}(t) = c_{2\omega_0}(t) = \sum_{n=0}^{\infty} \frac{4}{(2n+1)\pi} \sin{(2n+1)(2\omega_0)t},$$

which has spectral components only at the even harmonics $2\omega_0, 6\omega_0, 10\omega_0, \cdots$. Equation 4.6 now becomes

$$IQ'c_{2\omega_0}(t) + \delta_I Q'c_{90}(t) + \delta_Q Ic_0(t) + \delta_I\delta_Q + \delta_O, \tag{4.7}$$

and we achieve the goal of separating, in the frequency domain, the desired product from the artifacts of DC offsets.

The last steps are to do a down-chopping operation, and then to filter. For the down-chopping waveform, $c_{2\omega_0}(t)$ is the proper choice. To analyze the effect of multiplying expression 4.7 by $c_{2\omega_0}(t)$, it is helpful to make use of the equalities

$$\begin{aligned} c_0(t)c_{2\omega_0}(t) &= c_{90}(t) \\ c_{90}(t)c_{2\omega_0}(t) &= c_0(t) \\ c_{2\omega_0}(t)c_{2\omega_0}(t) &= 1, \end{aligned}$$

which are readily verifiable by the kind of graphical analysis depicted in figure 4.9. Multiplying expression 4.7 by $c_{2\omega_0}$, we obtain

$$IQ' + \delta_I Q'c_0(t) + \delta_Q Ic_{90}(t) + (\delta_I\delta_Q + \delta_O)c_{2\omega_0}(t). \tag{4.8}$$

Passing through a low pass filter at last yields the desired product, IQ'.

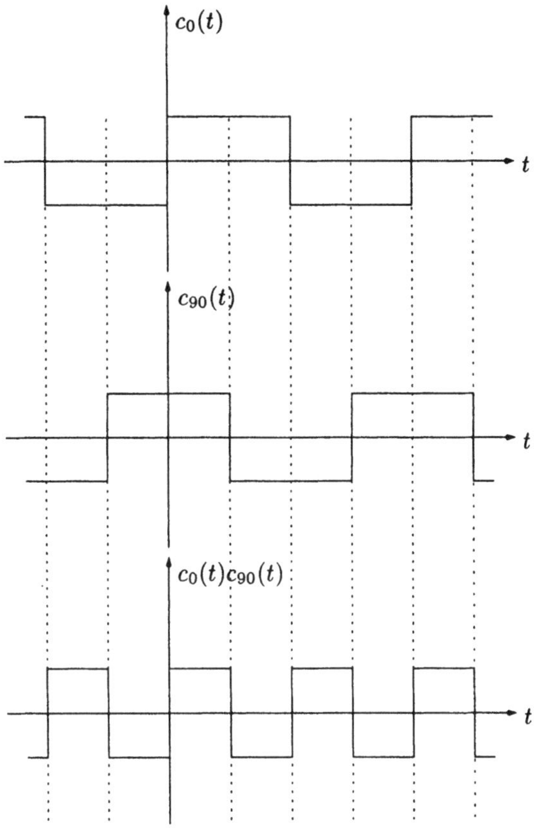

Figure 4.9. Graphically computing $c_0(t)c_{90}(t)$.

4.3.1 Limits on performance

Practical multipliers built using the chopping strategy of this section will enjoy substantially improved DC performance. The exact amount of improvement, however, is limited by the extent to which the mathematical abstractions of the foregoing analysis can be realized in practice. Chief among these abstractions are the ideal square waves as chopping waveforms, and the exact quadrature relationship between the two up-chopping clocks.

We are fortunate here that the gap between what is achievable in practice and what is demanded in theory is not prohibitively wide. CMOS technology, with its naturally available voltage switches, makes the chopping operation straightforward: simply commutating the two sides of a differential signal achieves the desired aim. Also, by using edge-triggered D-flip-flops, chopping clocks can be derived whose transitions occur only on the leading edge of a reference clock (see, for example, figure 5.17). To the extent that the reference clock is periodic, and regardless of its duty cycle, the chopping clocks will have a duty cycle of 50% and the quadrature relationship between the up-chopping clocks will be perfect.

4.4 Summary

This chapter essentially chronicles the evolution of a new phase alignment concept through two generations of development. The work begins by clearly setting down the mathematical consequences of phase misalignment, something that had not fully emerged in the literature. The first generation of development is represented by the original $IQ' - QI'$ controller, described in [42] and in appendix A. This discrete-component prototype not only serves well as a proof of concept, it also demonstrates the sensitivity of the controller to the DC offsets of the analog multipliers. In response, a new technique for realizing analog offset-free analog multiplication is proposed [43]. Inspired by chopper stabilization, it finds its first physical realization in the integrated circuit of chapter 5.

Chapter 5

A FULLY INTEGRATED CARTESIAN FEEDBACK SYSTEM

Among linearization techniques, Cartesian feedback has languished for many years behind predistortion and feedforward techniques in terms of widespread use. Even adaptive predistortion, carrying its overhead in power dissipation and complexity, has enjoyed more popularity in the literature. This chapter begins, then, with a brief examination of this state of affairs, and an explanation of the ideas that we explore with the CFB IC project. The rest of the chapter is devoted to describing the design of the prototype, and to the evaluation of its performance.

5.1 Motivation for pursuing Cartesian feedback

One naturally suspects that if Cartesian feedback has languished, it has done so for very good reasons. For a long time, there were two primary reasons. The first was that, being an analog feedback technique, it is necessarily bandwidth limited. The second was that there seemed no easy way to guarantee the phase alignment critical to system stability. After some initial investigation, however, the CFB IC project was started with the idea that these reasons were practical, not fundamental, and that the fundamental strengths of the technique justified a strenuous effort to overcome the practical problems.

Chief among the strengths is that compared to most other linearization techniques, the level of detail required in the power amplifier model is greatly reduced. The weight of this advantage cannot be overemphasized. At the time of this writing, power amplifier modeling remains a formidable task whose complexity shows no signs of diminishing. Even a casual browsing of the literature reveals an entire research community devoted to this very active field [44, 45, 46, 47, 48, 49, 50, 51, 52]. Linearization techniques that rely on the quality of the power amplifier model are therefore limited by that quality as well, as the path to better performance lies through the painstaking characterization of poorly understood phenomena.

The effectiveness of Cartesian feedback, on the other hand, is defined by the quality of the downconversion, the amount of loop gain, and the stability of the system. The limit imposed by the quality of the downconversion is fundamental to linearization strategies, as it is always necessary to look at the output baseband spectrum to evaluate performance. And if we allow for sufficiently conservative loop dynamics, stability and loop gain can be assured knowing only a worst-case group delay estimate of the power amplifier and bounds on its small-signal gain. For the trouble, a designer is rewarded with a system that is inherently low-power and resistant to drift.

In summary, there is strong theoretical motivation to pursue CFB. So it must be asked: can the practical drawbacks of the technique be overcome? As described in chapter 4, phase alignment seems suddenly to be manageable. But what about bandwidth, which is surely a fundamental limitation?

For that, we turn to figure 5.1 and observe that the signals at the baseband inputs of the upconversion mixers are *correctively distorted* versions of the desired baseband symbols. One can therefore view a Cartesian feedback system

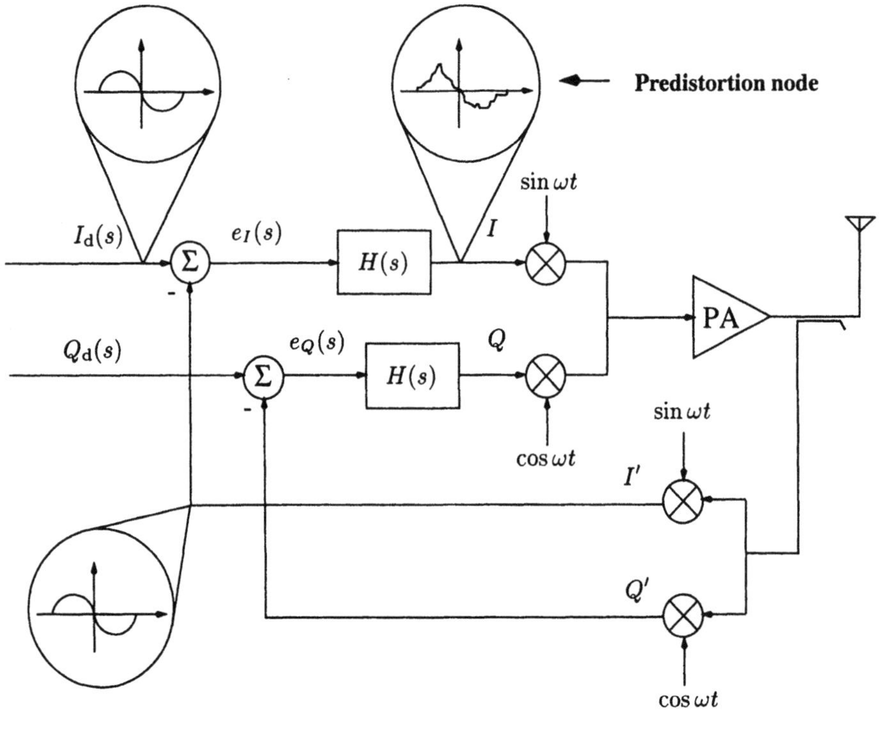

Figure 5.1. The predistorting action of Cartesian feedback.

as an analog computer, calculating at every instant the mixer input required to realize the desired baseband symbol. Suppose now that the transmitted data is drawn from a finite set of possible symbols, as is the case with any digital modulation system, and that the power amplifier, while nonlinear, is time-invariant. It follows that such a computer would need perform its calculation only once on a given symbol. If the result is then stored, it could be recalled upon the reappearance of the corresponding symbol.

The reader may recognize here the driving principle behind the traditional predistortion strategy. What predistortion famously lacks, however, is a way to deal with the fact that power amplifiers are not time-invariant. They age, for instance, and are subject to temperature and process variations. Adaptive predistortion is designed to take advantage of the relatively slow time scales of this time-variant behavior. Rather than have a continuously closed feedback loop, an adaptive predistorter updates its model as often as is deemed necessary. The result is that such a system does not suffer the bandwidth limitation that a Cartesian feedback system does.

In this dissertation a new use for Cartesian feedback is proposed, which is illustrated in figure 5.2. The heart of the idea is to use a Cartesian feedback

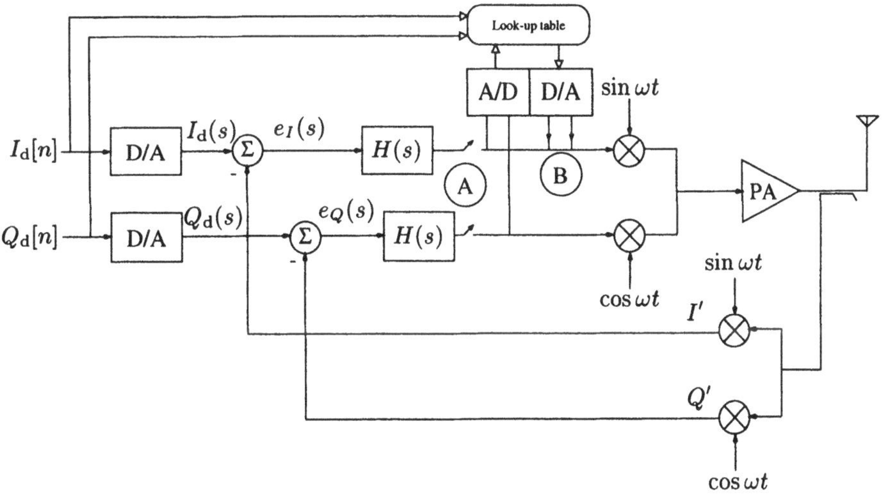

Figure 5.2. Cartesian feedback used to train a predistorter.

loop to train a predistorter. When the switches "A" in figure 5.2 are closed, the switches "B" are open and the system functions as an ordinary, closed-loop system. All of the symbols to be transmitted are stepped through in a suitable calibration sequence, with each symbol being held until the system settles completely, whereupon the output of the A/D converter is stored in a look-up table. Once this look-up table is complete, the switches "A" are opened, switches "B" are closed, and the system is run as an open-loop predistorter. When deemed necessary, the calibration sequence is done again.

In effect this is a lower power, lower complexity version of adaptive predistortion. It is lower power because there is no digital signal processor. Instead, a simpler finite-state machine for overseeing the calibration process suffices. It is lower complexity because no effort is made to maintain a detailed model of the power amplifier itself. One strategy in an adaptive predistorter, for example, is to model the power amplifier transfer characteristic with a polynomial of many terms, and then update the coefficients of those terms to minimize distortion products[22]. In part, the complexity lies in trying to update the model in a way that successfully and quickly converges. One has traded the analog stability problem (of Cartesian feedback) for a nonlinear, multivariate one.

The main advantage of the idea is that calibration can be done at low symbol rates, where loop gain is plentiful, and then actual transmission can be done at high symbol rates while not constrained by the speed of the closed-loop system. Thus the bandwidth limitation, long a liability of Cartesian feedback, is effectively circumvented.

A concern at the beginning of this project, however, was the possibility that data gathered at low symbol rates is no longer valid at high symbol rates. Researching the literature quickly quieted these fears. Open-loop predistorters are regularly tuned using one or two unmodulated carrier tones[20, 19]. The case of one tone, in particular, corresponds to a static characterization from the standpoint of baseband input. The open-loop predistortion literature thus represents a body of empirical evidence that the nonlinear characteristics of power amplifiers do not change significantly over typical baseband bandwidths.

At this point, having found ways to deal with the bandwidth limitation and the phase alignment problem, it was decided to implement a prototype Cartesian feedback system as a platform for exploring these ideas. The next major decision was whether to design an IC, or to build a discrete-component system.

5.2 Motivation for a monolithic implementation

For this project a monolithic implementation and a discrete-component implementation are conceptually equivalent. The decision to build an IC was based primarily on industry interest. Had a discrete version been successfully built, questions concerning the feasibility of a fully integrated version were anticipated. The hope was that by demonstrating, on a single die, a power amplifier, a phase alignment system, and a complete Cartesian feedback loop, this project might gain a new relevance for Cartesian feedback in this era of highly integrated solutions.

There are some aspects of the design that are genuinely easier for a monolithic design. Above all, a same-chip implementation allows for minimum power consumption in the linearization circuitry. To see why, consider the proliferation of analog multipliers in this system. Discrete multipliers, of the type used in the phase alignment prototype, are often designed to achieve high bandwidth, 1-volt output swings while driving resistive loads as low as 25 ohms. While the speed is certainly useful, the ability to drive resistive, off-chip loads translates into large power dissipation requirements. The on-chip environment, where output loads for circuit blocks are no larger than the gates of a few MOSFETs, can be exploited to achieve the lowest possible power dissipation for the linearization and phase alignment circuitry.

The prototype Cartesian feedback IC is implemented in National's 0.25μm process, and operates with a 2.5V supply.

5.3 CFB IC at the system level

Before the individual circuit blocks can be designed, a few general considerations must be addressed. Chief among these is isolation of the linearization circuitry from the large-signal, high-frequency signal swings of the power amplifier. A known liability of circuits implemented in a CMOS process, this possibility of substrate coupling is dealt with by making all signal paths, to the extent possible, fully differential. Additional confidence derives from the disparate frequency bands in which the different subsystems operate. The power amplifier operates at 2 GHz, while the linearization circuitry, based as it is on op-amp-type topologies, is largely insensible to signal frequencies greater than 10 MHz or so.

Otherwise, the following list details five of the most important overall considerations for the design:

- **Downconversion mixers:** the system should have the highest-quality downconversion mixer possible. While mixers have many performance measures, linearity is paramount for this system.

- **Predistort I/O:** access to the "predistortion nodes" of figure 5.1 must be provided to the user. A way must be provided for switching between driving the upconversion mixer directly and ordinary, closed-loop operation.

- **Adjustable loop dynamics:** the user must be able to alter the transfer function, $H(s)$, of figure 5.1. This will allow the chip to be a vehicle for exploring different loop compensation strategies in a Cartesian feedback system.

- **Managing DC offsets:** within various subsystems, it happens that successive stages, either performing amplification or some other analog signal processing function, are DC coupled. It is important to ensure that DC offsets alone never cause saturation.

- **Testing for phase alignment:** for characterizing the tracking behavior of the phase alignment system, there must be a means of manually introducing phase misalignment.

Figure 5.3 is a conceptual diagram of the complete integrated circuit that serves to demonstrate many of the ideas developed in this dissertation. Discussion

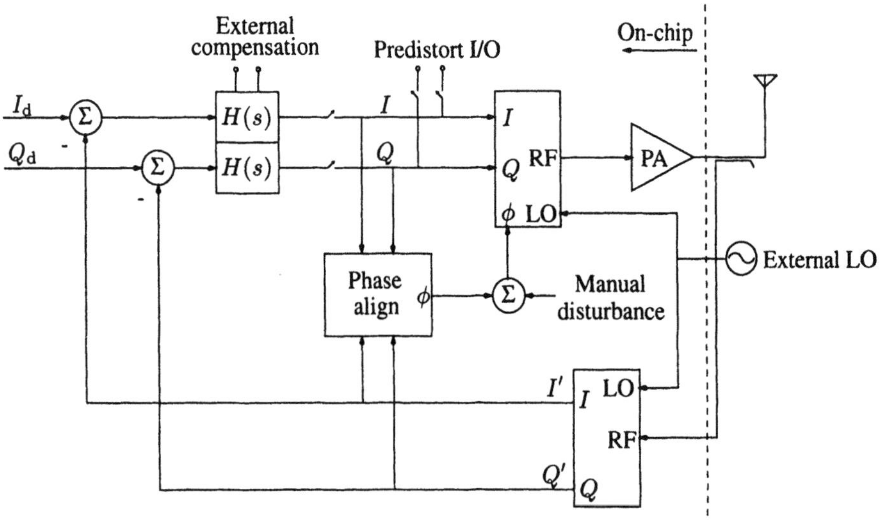

Figure 5.3. Conceptual diagram of CFB IC.

of the circuit details proceeds in two main parts. Section 5.4 is a detailed description of implementation of the phase alignment system, while section 5.5 covers the remainder of the IC.

5.4 The phase alignment system

As seen in chapter 4, the mathematical description $\frac{d\theta}{dt} = G(IQ' - QI') = -Grr' \sin(\theta - \theta')$ of the phase alignment concept provides, for the most part, a straightforward mapping onto a circuit realization. The time derivative implies the presence of an integrator, the sum of products points to two analog multipliers and subtraction, and the factor G implies amplification. What is *not* obvious from the mathematics is how to actually realize the necessary phase shift. While in principle the corrective phase shift can be done at either the upconversion or downconversion stage, in practice it should be done at the upconversion system block. The reason for this recommendation is that it is impossible not to introduce some noise or distortion by the very act of rotation. The upconversion block is in the forward path of the Cartesian feedback loop, and so here the added signal corruption is suppressed.

Otherwise, the designer has considerable freedom in realizing the physical implementation. Figures 5.4 and 5.5 show two mathematically equivalent methods of phase shifting for the Cartesian feedback IC. Actually phase shifting

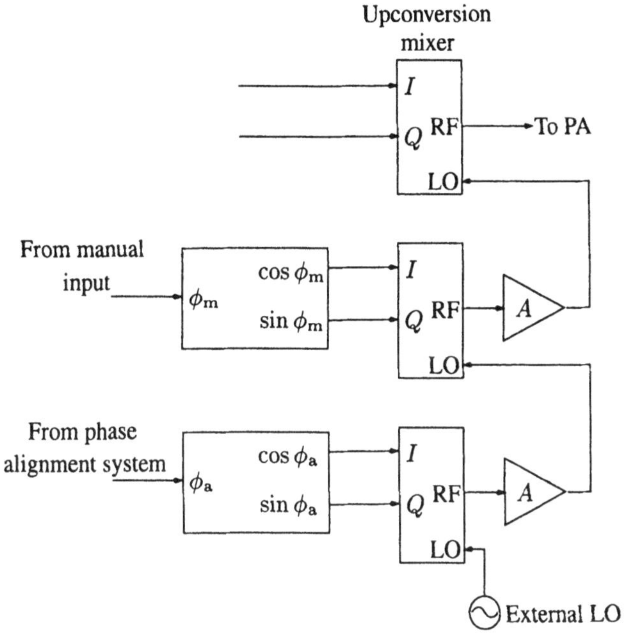

Figure 5.4. Phase alignment by phase shifting the local oscillator.

the local oscillator, as shown in figure 5.4, is conceptually the most obvious choice, and is the method used in the original discrete-component prototype. The phase shifters themselves are simply quadrature modulators with appropriate control signals fed into the IF ports, an implementation amenable to integration. Phase shifting the local oscillator has the practical shortcoming of amplifiers A, necessary because the LO port of a quadrature modulator often requires a very high amplitude signal. It follows that at multi-gigahertz carrier frequencies, the power overhead incurred by generating successive LO signals can be considerable.

Rotation of the baseband symbol is thus a lower power solution, and is used for the prototype IC. Shown in figure 5.5 are two analog matrix multiplication blocks. Four analog multipliers form the core of each.

Not shown in either figure 5.4 or figure 5.5 are the details of the blocks that take ϕ as an input and produce $\sin\phi$ and $\cos\phi$, or signals proportional thereto, as outputs. No such blocks to produce this literal transformation are used for this work. Rather, a simplifying compromise is used that still accomplishes the desired end. Mathematically, a first step is to feed the voltage input, v_ϕ, directly

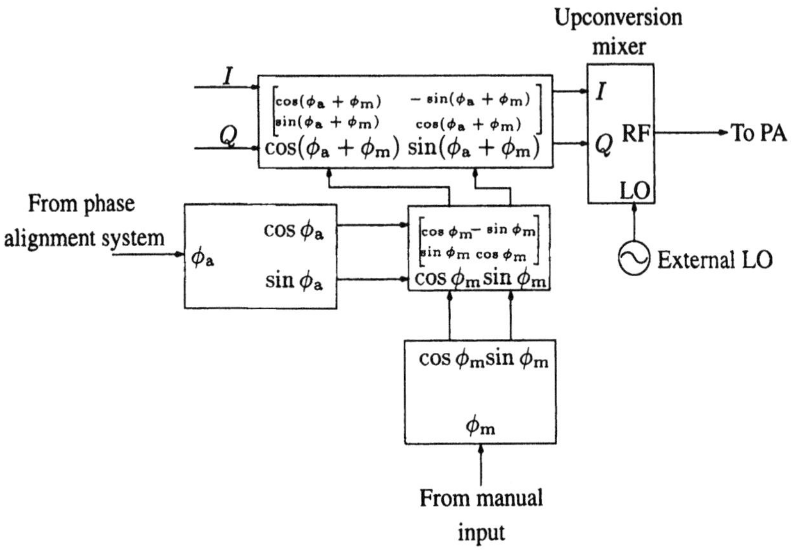

Figure 5.5. Phase alignment by rotating the baseband symbol.

through to one of the outputs

$$v_{\sin\phi} = v_\phi.$$

Keeping in mind the relation $\sin^2\phi + \cos^2\phi = 1$, the goal is achieved if it is arranged that

$$v_{\cos\phi} = \sqrt{K - v_{\sin\phi}^2},$$

where K is a constant. Figure 5.6 is a conceptual diagram of the analog circuit used in the original prototype. Generating $v_{\sin\phi}$ and $v_{\cos\phi}$ this way, as opposed to a literal deriving from an input v_ϕ, has minor mathematical consequences, a brief discussion of which can be found in section A.1.

The concept of figure 5.6 has the advantage of realizing a pure rotation without an angle-dependent warping of the symbol magnitude. It suffers the circuit liability of requiring analog squarers, which in practice are difficult to build without introducing substantial DC offsets. Before forging ahead with the IC prototype, then, another simplification is needed.

Consider the solution shown in figure 5.7. Here the control loop acts to preserve the 1-norm of the $(v_{\sin\phi}, v_{\cos\phi})$ pair: $|v_{\sin\phi}| + |v_{\sin\phi}| = K$. For $|v_\phi| \leq K$, rotations over a range of ± 90 degrees can be achieved. The analog squarers are gone, and in their place we have the potential complication of a folding amplifier to realize the absolute value function. Fortunately the

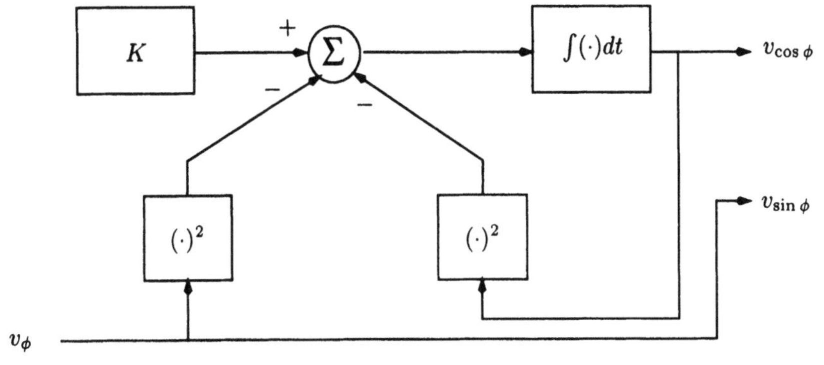

Figure 5.6. Analog technique for generating $v_{\sin\phi}$ and $v_{\cos\phi}$.

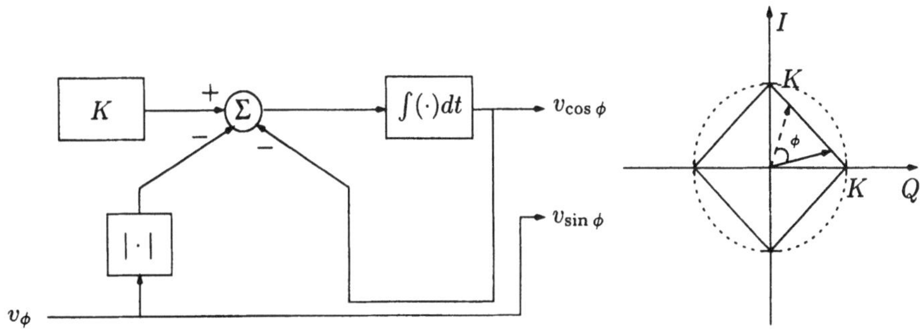

Figure 5.7. Analog rotation using the 1-norm.

offset-free voltage switches, or pass gates, available in CMOS enable a simple implementation strategy for this block if the signal paths are differential. One idea is illustrated in figure 5.8, which shows a bridge of four switches between the input and the output. A comparator senses the sign of the input, and configures the switches such that the output is always of the same sign.

This simplicity of implementation comes at a price, which is that now the symbol magnitude *is* warped as a function of the rotation angle. Specifically, figure 5.7 shows that over the course of any continuous 90-degree rotation, the minimum magnitude is factor of $\frac{\sqrt{2}}{2} \approx 0.707$ smaller than the maximum magnitude. By performing the rotation on the upconverted signal, however, this distortion is introduced in the forward path of the Cartesian feedback system, where it is maximally suppressed.

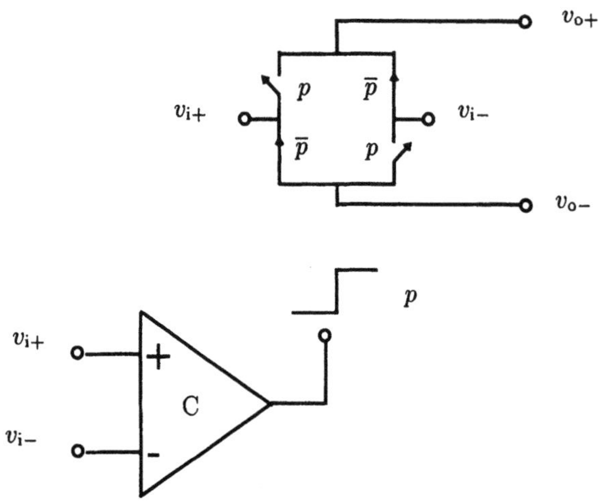

Figure 5.8. Using CMOS voltage switches and a comparator to realize a folding amplifier. Switches are closed when their respective control signal is high.

5.4.1 Circuit details

The list of necessary circuit blocks follows quickly and naturally from the conceptual framework established in chapter 4 and thus far in section 5.4: a basic multiplier cell, a phase error computation block, an analog integrator, a circuit realization of figure 5.7, and an analog matrix rotation block. The following paragraphs detail the design of the circuit blocks used in the phase alignment portion of the CFB IC.

5.4.1.1 Basic multiplier cell

The topology for the multiplier cell, shown in figure 5.9, is chosen from among the many possibilities that have been documented for CMOS technology [39]. Conceptually, the circuit can be understood by looking at the output current contributed by the first numbered, stacked transistor pair. The bias voltages X and Y are chosen such that the bottom transistor operates in the triode region. The top transistor functions as a source follower, or voltage buffer, and sets V_{ds} of the bottom transistor. The underlying principle then becomes clear from the drain current expression for a transistor operating in triode,

$$
\begin{aligned}
I_{\mathrm{d}} &= \mu_{\mathrm{n}} C_{\mathrm{ox}} \left(\frac{W}{L}\right) \left((V_{\mathrm{gs}} - V_{\mathrm{T}}) - \frac{V_{\mathrm{ds}}}{2}\right) V_{\mathrm{ds}} \qquad (5.1)\\
&= \mu_{\mathrm{n}} C_{\mathrm{ox}} \left(\frac{W}{L}\right) \left(V_{\mathrm{gs}} V_{\mathrm{ds}} - V_{\mathrm{T}} V_{\mathrm{ds}} - \frac{V_{\mathrm{ds}}^2}{2}\right).
\end{aligned}
$$

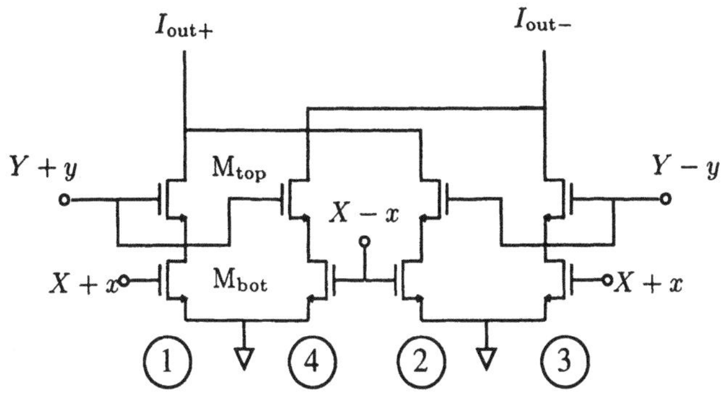

Figure 5.9. Basic topology for multiplier cell. All transistors connected to a y input are sized M_{top}, and all transistors connected to a x input are sized M_{bot}.

The term $V_{gs}V_{ds}$ is what we want, since V_{ds} is set by y through the source follower and V_{gs} is set by x directly. In the following we show that the other three stacked pairs serve primarily to cancel the undesirable terms in equation 5.1.

For the detailed derivation, we start by writing for I_1, the current output of transistor pair 1:

$$I_1 = \mu_{\text{n}} C_{\text{ox}} \left(\frac{W}{L}\right)_{\text{bot}} \left(X + x - V_{\text{T}} - \frac{Y - V_{\text{GS-top}} + y}{2}\right)(Y - V_{\text{GS-top}} + y)\,, \tag{5.2}$$

where the V_{T} is the threshold voltage (body effect is ignored), and $V_{\text{GS-top}}$ is the bias value gate-to-source voltage of the top transistors. It is convenient to expand equation 5.2 slightly:

$$\begin{aligned} I_1 \;=\;& \mu_{\text{n}} C_{\text{ox}} \left(\frac{W}{L}\right)_{\text{bot}} ((X - V_{\text{T}})(Y + y - V_{\text{GS-top}}) + xy + x(Y - V_{\text{GS-top}}) - \\ & \left.\frac{(Y - V_{\text{GS-top}} + y)^2}{2}\right). \end{aligned}$$

Now, do the same for the other stacked pairs:

$$\begin{aligned} I_2 \;=\;& \mu_{\text{n}} C_{\text{ox}} \left(\frac{W}{L}\right)_{\text{bot}} ((X - V_{\text{T}})(Y - y - V_{\text{GS-top}}) + xy - x(Y - V_{\text{GS-top}}) - \\ & \left.\frac{(Y - V_{\text{GS-top}} - y)^2}{2}\right), \end{aligned}$$

$$I_3 = \mu_n C_{ox} \left(\frac{W}{L}\right)_{bot} \left((X - V_T)(Y - y - V_{GS-top}) - xy + x(Y - V_{GS-top}) - \frac{(Y - V_{GS-top} - y)^2}{2}\right),$$

and

$$I_4 = \mu_n C_{ox} \left(\frac{W}{L}\right)_{bot} \left((X - V_T)(Y + y - V_{GS-top}) - xy - x(Y - V_{GS-top}) - \frac{(Y - V_{GS-top} + y)^2}{2}\right).$$

The last step is to combine terms according to

$$I_{out} = I_{out+} - I_{out-} = (I_1 + I_2) - (I_3 + I_4).$$

Doing so results in the output current

$$I_{out} = \mu_n C_{ox} \left(\frac{W}{L}\right)_{bot} (4xy).$$

It is important to emphasize that the foregoing derivation hinges on approximating M_{top} as a source follower, allowing us to treat V_{GS-top} as a constant offset independent of y. To assure that this approximation is valid in practice, we turn to the expression for the voltage output of an ordinary source follower,

$$V_{out} = \frac{g_m R_s}{1 + g_m R_s} V_{in}, \tag{5.3}$$

where R_s in equation 5.3 is the resistance in the source. The goal of a voltage buffer is achieved to the extent that we ensure $g_m R_s \gg 1$. For the multiplier, then, we require $\frac{g_{m-top}}{g_{ds-bot}} \gg 1$. The drain-source conductance of the bottom transistor varies greatly with V_{DS-bot}, but we can establish the upper bound

$$g_{ds-bot}|_{V_{DS-bot}=0} = \mu_n C_{ox} \left(\frac{W}{L}\right)_{bot} (V_{GS-bot} - V_T).$$

Recalling that

$$g_{m-top} = \mu_n C_{ox} \left(\frac{W}{L}\right)_{top} (V_{GS-top} - V_T),$$

we require for the multiplier

$$\frac{g_{m-top}}{g_{ds-bot}} = \frac{\left(\frac{W}{L}\right)_{top} (V_{GS-top} - V_T)}{\left(\frac{W}{L}\right)_{bot} (V_{GS-bot} - V_T)} \gg 1. \tag{5.4}$$

This condition is satisfied in practice by appropriate transistor sizing.

Figure 5.10 shows the actual multiplier cell, which differs from figure 5.9 only by the presence of transistors M9-M15. The system blocks that use

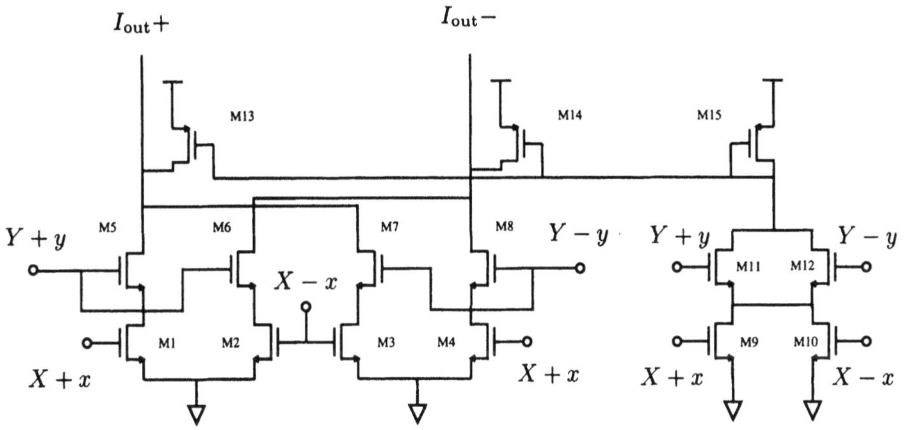

Figure 5.10. Multiplier cell.

Device	Value
M1-M4	4/10
M5-M8	20/5
M9-M10	2/40
M11-M12	10/20
M13-M14	40/0.5
M15	10/0.5

Table 5.1. Multiplier elements.

the output of this multiplier require voltage input, and so the drains of devices M5-M8 are connected to resistive loads to perform an I-V conversion. Without these extra devices, the resistive loads set both the bias point of the following (DC coupled) amplifier stage as well as the voltage gain of the multiplier. It is convenient to decouple these two problems. M9-M15 provide some of the drain current for transistors M5-M8, and thereby reduce the bias current flowing through the resistive load. By sizing M15 relative to M13 and M14, the designer is free to vary the output bias voltage and the multiplier gain with some degree of independence.

Finally, note that actual use of this multiplier cell requires the two input signals to be at *different* common-mode levels. The common mode level X should be higher than Y in order to ensure that the bottom devices stay in the triode region. The default common-mode level for differential signals in this system is chosen to be mid-supply (1.25V). Where this multiplier is used, then, active steps are required to ensure that X is at least 500mV above mid-supply.

5.4.1.2 Phase error computation

The phase error computation block is built of three simple elements: analog multipliers (as described in section 5.4.1.1), commutating mixers, and an amplifier. The commutating mixers, or "choppers," are all identical, and a schematic is shown in figure 5.11. The sizing of these devices is not critical.

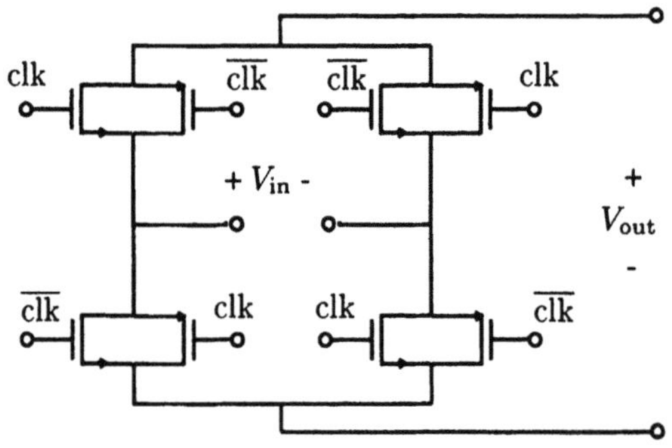

Figure 5.11. Commutating mixer for chopping. NMOS devices are sized 3/0.24, PMOS 9/0.24

The choppers are designed to operate at a maximum of 5MHz when driving the light capacitive loads of a few MOSFET gates. The only other guideline is that the switches be made small so as to minimize the capacitive coupling from the clocks to the output.

Device	Value
M1-M3	12/1
M4-M7	3.3/0.5
M8	49/1
M9-M10	10/2
M11-M14	4/4
Quiescent current	406μA

Table 5.2. Phase error computation elements. Quiescent current includes current draw of multiplier cells.

The multipliers are the cells described in section 5.4.1.1, and the required subtraction is carried out in the current domain. The current-to-voltage conversion is accomplished via the resistive loads M9-M10, and two amplification

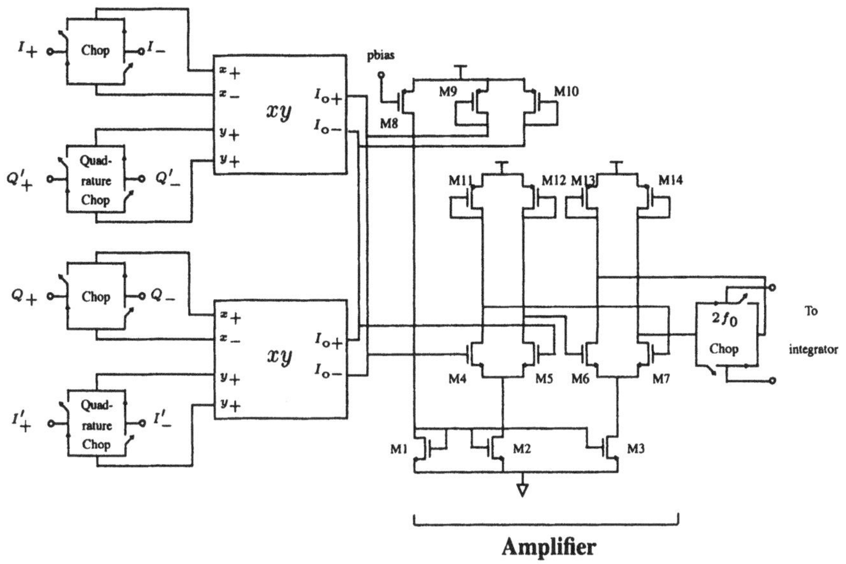

Figure 5.12. Phase error computation.

stages immediately follow. The purpose of these stages is to provide AC amplification of the product centered at $2f_0$. In many chopper-stabilized schemes this would be a true AC amplifier with capacitively coupled inputs. On-chip capacitive coupling, however, is inconvenient on at least two counts. First, sufficiently large coupling capacitors occupy a great deal of area. Second, the substantial bottom plate parasitic of on-chip capacitors incurs an unavoidable performance degradation. As an overall design strategy, AC coupling is avoided except where absolutely necessary.

5.4.1.3 An analog integrator

A switched-capacitor (S.C.) integrator drives the realization of the phase alignment concept, and two factors motivate this choice over continuous-time methods. The first factor is that the offset removal accomplished by chopper stabilizing the multipliers is wasted if the integrator that follows has a large input-referred offset. It follows that an autozeroing integrator, of the type commonly realized via switched-capacitor techniques,[1] is an attractive option [53]. The second factor is that high speed in the phase alignment system is unnecessary, as the proper phase setting typically evolves on time scales no shorter than those of temperature change, aging, and process variation. A slow

[1] Such integrators go by various descriptors in the literature. The integrator detailed in this section, for example, is sometimes called a "correlated double sampling integrator."

integrator is thus appropriate, and easily realized with a slowly clocked S.C. integrator.

The design of a S.C. integrator starts with the design of a fully differential op-amp. A gain-boosted, folded cascode topology is selected from among the many known choices [54, 55, 56], mainly because of its high DC gain and relatively uncluttered dynamics. Figure 5.13 is a schematic of the final op-

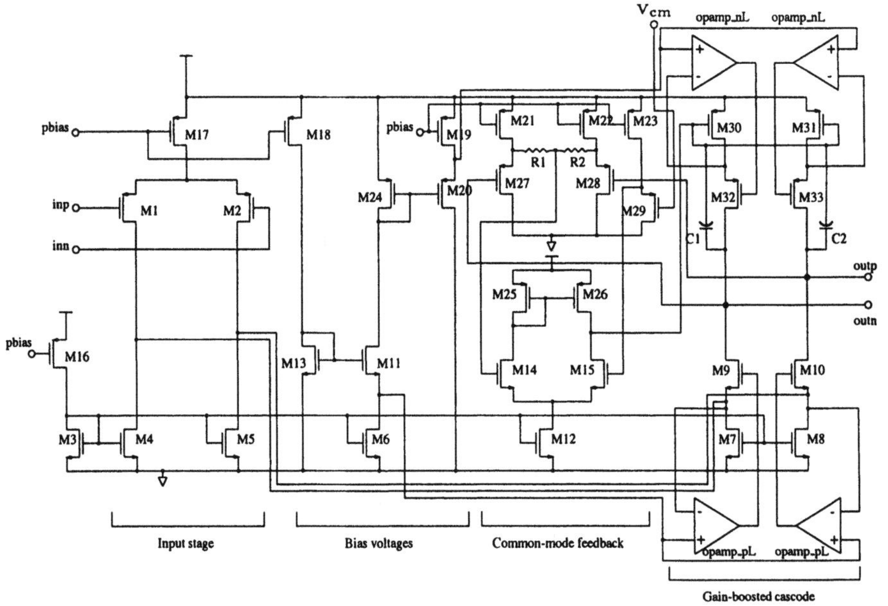

Figure 5.13. Op-amp_d1, a fully differential op-amp for the S.C. integrator.

Device	Value	Device	Value
M1-M2	2×40/1.5	M25-M26	20/1
M3-M11	25/1	M27-M29	11/0.24
M12	12.5/1	M30-M31	2×49/1
M13	5/1	M32-M33	49/1
M14-M15	6/3	C1-C2	2.5pF
M16-M23	49/1	R1-R2	60kΩ
M24	9/1		
Quiescent current	524μA		

Table 5.3. Elements for integrator op-amp. Quiescent current includes current draw of Opamp_nL and Opamp_pL.

amp. Care is taken to minimize the V_T mismatch between devices in the input differential pair. They are large transistors, and each is split into two parts to allow a common-centroid layout. The section labeled "Bias voltages" in the schematic is a network to establish bias voltages for the cascode stack, and device sizes are chosen to maximize output voltage swing. The common-mode feedback (CMFB) is handled by a single-ended, two-stage op-amp, which sets the common-mode level to match the V_{cm} input. Capacitors C1 and C2 serve as the Miller-multiplied compensation capacitors for the CMFB circuit, and have the additional benefit of stabilizing the overall, differential op-amp. The unloaded differential unity gain frequency of this op-amp is 17MHz, with a DC gain of 115dB. The CMFB circuit demonstrates an 89 degree phase margin even with a 3pF capacitive load on each differential output.

The schematics of the single-ended op-amps that perform the gain boosting in figure 5.13 are shown in figures 5.14 and 5.15. As employed in figure 5.13, the inputs to these op-amps are going to rest within a V_{dsat} of a supply rail. It is thus necessary to design two varieties. One, shown in figure 5.14, is designed to handle inputs close to ground. The other, shown in figure 5.15, is designed to handle inputs close the positive supply rail. These amplifiers play general purpose roles. Accordingly, capacitance M18 in both circuits is placed to ensure stability regardless of the capacitive load being driven. The folded-cascode topology means that stability will be improved as the capacitance of the load increases, and in this case it is actually necessary to provide a minimum capacitance to ensure stability in the unloaded case.[2] As shown, both op-amps achieve a worst-case phase margin of better than 62 degrees.

Device	Value
M1-M2	2×7/0.6
M3-M9	10/1
M10-M11	49/1
M12	3/2
M13	24.5/1
M14-M17	25/1
M18	10/9
Quiescent current	75.9μA

Table 5.4. Opamp_pL elements.

With the differential op-amp established, the core of the integrator is in place. The rest of the design is determined by larger system issues. Chief among these issues concerns testability: it would be nice to be able to "turn off" the phase alignment system, or have it sit by dormant while allowing the phase alignment to be adjusted manually. In the original prototype the solution is to implement

[2]Stability margins are checked with op-amps configured as unity-gain buffers.

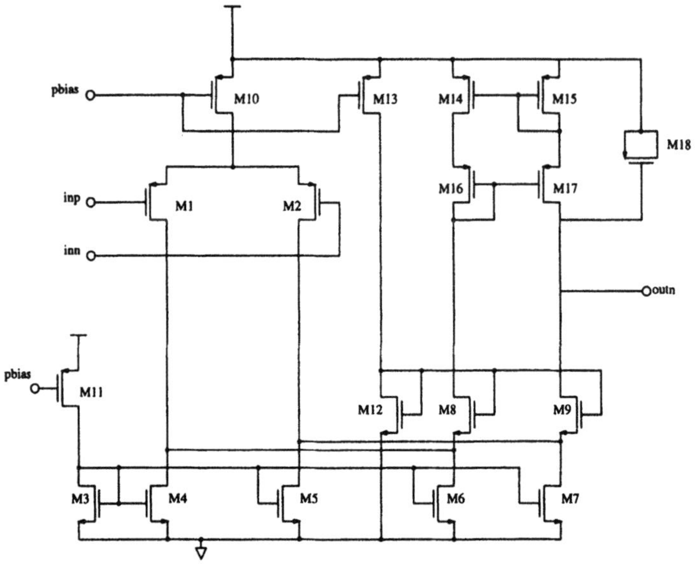

Figure 5.14. Opamp_pL, a single-ended op-amp for low common-mode inputs.

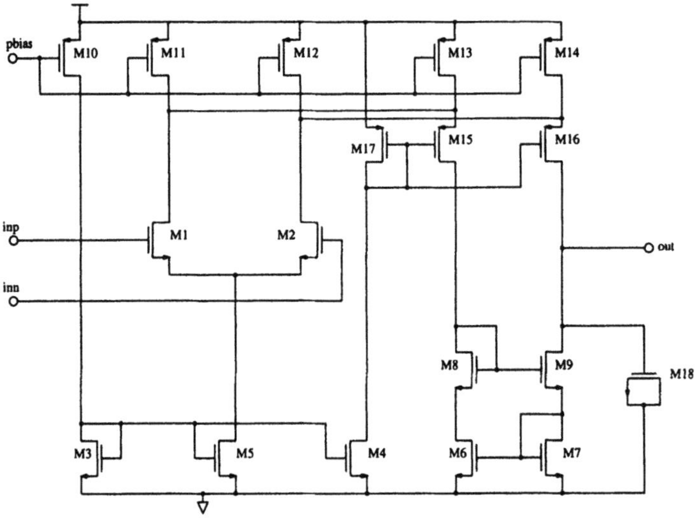

Figure 5.15. Opamp_nL, a single-ended op-amp for high common-mode inputs.

a 3-mode continuous-time integrator, the three modes being preset, hold, and normal integration. For this S.C. integrator it is convenient to implement two modes. The first mode is simply normal integration. For the second mode the inverted output of the integrator is connected to the input, with the result that

Device	Value
M1-M2	2×3.5/1
M3-M4	10/1
M5	2×10/1
M6-M9	5/1
M10-M16	25/1
M17	6/1
M18	11/11
Quiescent current	63.9μA

Table 5.5. Opamp_nL elements.

the output is eventually driven to zero. Use of this second, reset mode during testing enabled verification of the manual rotation system, as well as numerous other experiments.

Figure 5.16 is a diagram of the autozeroing, switched-capacitor integrator. The integrator relies on two non-overlapping clock phases. During clock

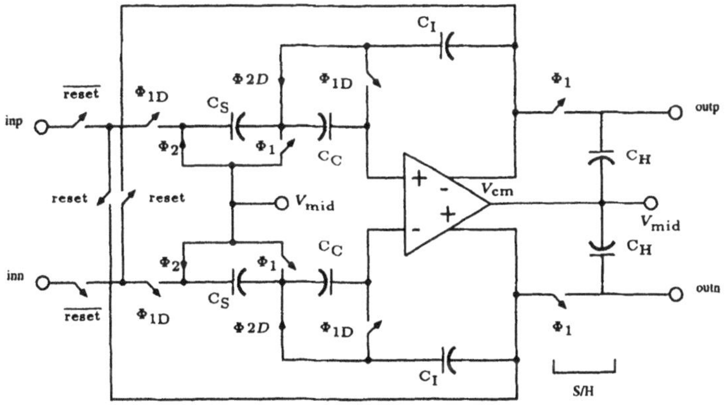

Figure 5.16. Switched-capacitor, non-inverting integrator for phase alignment system. Switches are complementary: NMOS 2/0.24 , PMOS 6/0.24 .

Device	Value
C_S	0.2pF
C_C	0.2pF
C_I	2pF
C_H	1pF

Table 5.6. Integrator capacitor values.

phase 1, the input is sampled and stored on capacitors C_S, and the input offset of the op-amp is stored on capacitors C_I. Integration occurs on clock phase 2.

Use of this sampled-data integrator for continuous-time control requires a minor adaptation. Even assuming zero input and perfect cancellation of the input-referred offset, the output exhibits a jump in voltage each time the top plates of the integration capacitors C_I are switched from one side of C_C to the other. The magnitude of the jump equals the magnitude of the differential voltage stored on capacitors C_C, which is just the magnitude of the op-amp's input-referred offset. Unless a corrective step is taken, these jumps amount to an undesired square wave at the sampling frequency added to the output of the op-amp. As shown in figure 5.16 the solution for this IC is to add a sample-and-hold (S/H). The output of the op-amp is sampled on clock phase 1 and held on phase 2, with the result that the overall circuit better approximates a continuous-time integrator.

Bottom plate sampling is used to reduce signal-dependent corruptions due to switch charge injection. The general principle is illustrated by examining the switching of the capacitors C_S. Φ_1 and Φ_{1D}, a delayed version of Φ_1, are high during the sampling period. At the end of the sampling period, Φ_1 releases the bottom plates of C_S *first*, leaving part of the channel charge in these Φ_1 switches on C_S. But because one side of these switches is tied to a low impedance source, most of the channel charge is distributed there and away from the sampling capacitors. Moreover, in principle the charge that gets injected onto C_S is the same every time. This is because the conditions of the switches are the same after every sampling instant: the source and drain rest at V_{mid}, regardless of the signal level. When Φ_{1D} then releases the top plates, the bottom plates are already floating and no charge can be injected from the Φ_{1D} switches.

To realize a slow integration, the sampling clock is in the range of tens of kilohertz. The chopping frequency f_0 for the multipliers is 2.5MHz, and down-chopping occurs at 5MHz. Aliasing of chopping artifacts is thus a concern. Filtering is undesirable because it complicates the dynamics of the phase alignment control system. Instead, the integrator is clocked at 39.2kHz, which has the property of aliasing tones at 2.5MHz and 5MHz to $\frac{\pi}{4}$ and $\frac{\pi}{2}$, respectively, in the z-domain.[3] As a result, aliasing of these artifacts does not result in DC errors. All clocks are derived from a 20MHz off-chip crystal oscillator. Chopping clocks transition on the rising edge of the 20MHz source, while the integrator clock transitions on the falling edge. This ensures that edges of the chopping clocks do not occur at a sampling instant.

[3] Any frequency determined by $\frac{4f_0}{n}$, where n is an odd integer, will have this property. For this chip, n is 255.

Figure 5.17 shows the straightforward way in which the chopping clocks are derived from the 20MHz source. Two D flip-flops ensure a quadrature

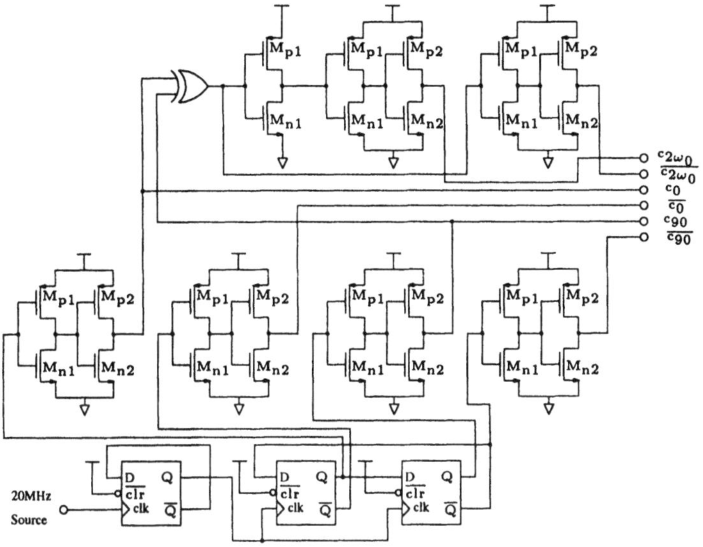

Figure 5.17. Chopping clocks derived from off-chip source.

Device	Value
M_{n1}	2/0.24
M_{n2}	4/0.24
M_{p1}	6/0.24
M_{p2}	12/0.24

Table 5.7. Elements for chopping clocks.

relationship between c_0 and c_{90}. The clock for down-chopping is derived from c_0 and c_{90} using an XOR gate. It is important to note that the designation of $c_{2\omega_0}$ versus $\overline{c_{2\omega_0}}$ is *not* arbitrary, as exchanging the two inverts the sign of the final phase error signal.

The circuit for generating the integrator's clock signal is shown in figure 5.18. The requirements for this clock signal are that it transition on the trailing edge of the external reference, and that its frequency be related to f_0 by $\frac{4f_0}{n}$, where n is an odd integer. The first requirement is satisfied by the inclusion of the inverter between the external reference and the clock input of the ripple counter. The second is satisfied by dividing the 20MHz reference by 2×255, as shown in the figure. Since 20MHz is eight times faster than f_0, the $\frac{4f_0}{n}$ requirement is

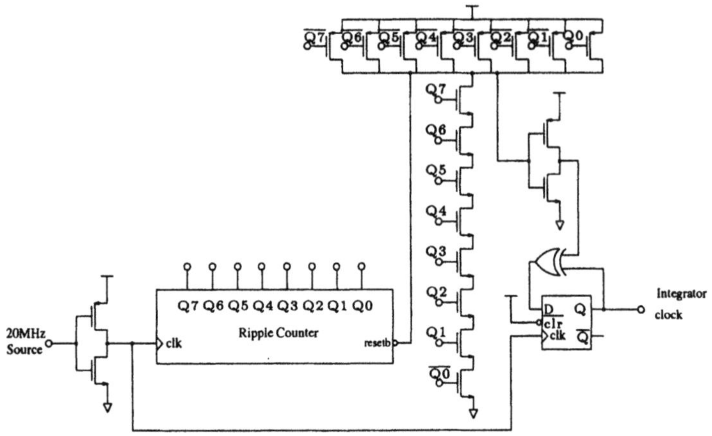

Figure 5.18. Integrator clock, which transitions on the trailing edge of the external 20MHz source.

met with n equal to 255. Finally, non-overlapping clock phases are generated from this integrator clock signal with the circuit shown in figure 5.19 [57].

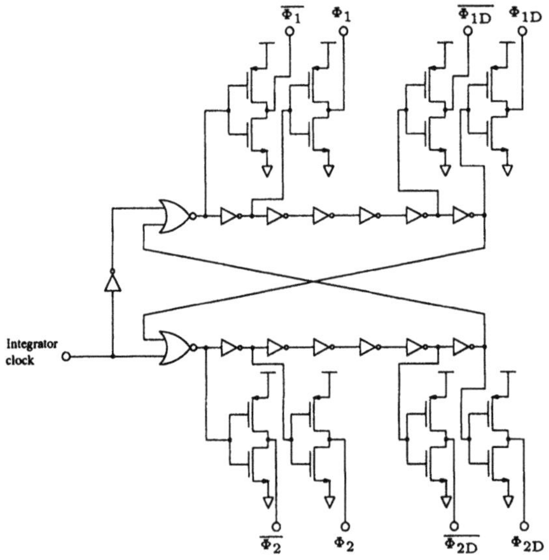

Figure 5.19. Circuit for generating clock phases.

5.4.1.4 A constant 1-norm controller

To perform symbol rotation, this phase alignment system implementation requires a circuit realization of figure 5.7. The path from concept to circuit is particularly straightforward if the necessary subtractions are carried out in current domain, as shown in figure 5.20.

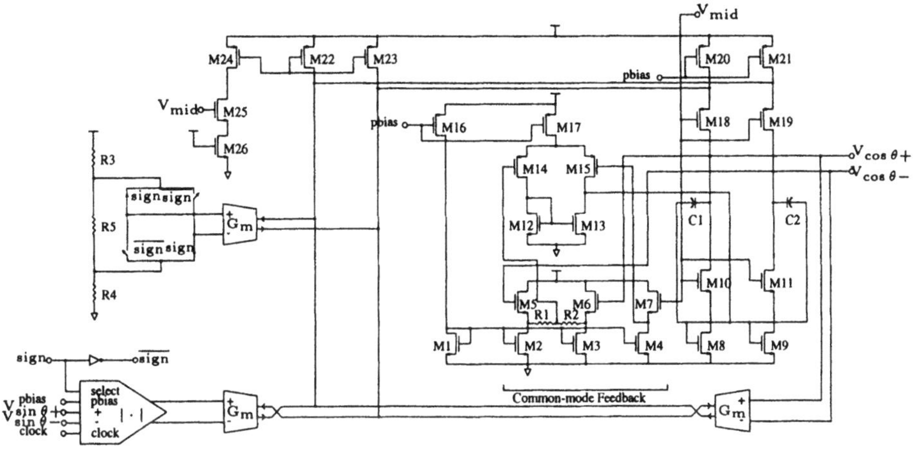

Figure 5.20. Constant 1-norm controller: circuit realization of figure 5.7.

Device	Value	Device	Value
M1-M4	12/1	M20-M21	25/1
M5-M7	4/0.24	M22-M23	6×20/1
M8-M9	25/2	M24	20/1
M10-M11	12/2	M25	5/12
M12-M13	12/1	M26	1/20
M14-M15	11/8	C1-C2	1.3pF
M16	49/1	R1-R2	30kΩ
M17	2×49/1	R3-R4	4×3.33kΩ
M18-M19	49/2	R5	2×3.33kΩ
Quiescent current	400μA		

Table 5.8. Constant 1-norm elements. Quiescent current includes current draw of folding amplifier and G_m cells.

As with the fully differential op-amp of this section, the common-mode output level is fixed to mid-supply via a two-stage op-amp. The first stage is transistors M12-M17, and capacitors C1-C2 serve both as the Miller capacitance for the common-mode feedback loop and to stabilize the overall, fully-differential 1-norm controller. Driving a capacitive load of 500fF on each differential output, this circuit demonstrates a closed-loop bandwidth of 1.3MHz.

A feature of the constant 1-norm circuit of figure 5.20 is that an external signal, *sign*, determines the sign of $V_{\cos\theta}$. The voltage $V_{\cos\theta}$ is positive when *sign* is high, and negative when *sign* is low. This feature proved useful in testing by extending the range of possible manual rotations from 180 to a full 360 degrees (see figure 5.5).

Figure 5.21 depicts the transconductance cell labeled 'G_m' in figure 5.20. A differential voltage applied to the gates of M2 and M4 results a current

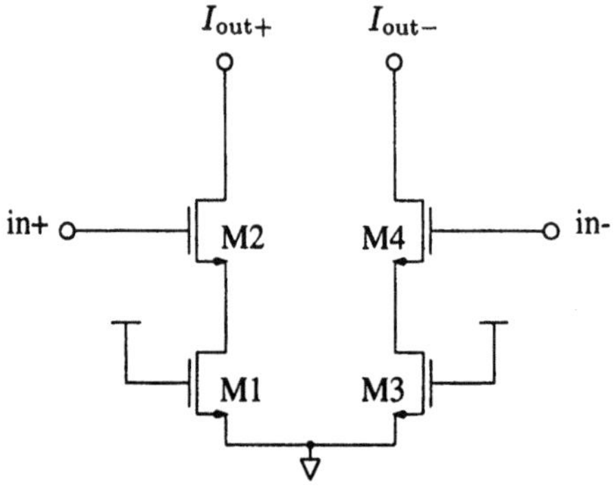

Figure 5.21. Differential transconductor.

Device	Value
M1, M3	2×1/20
M2, M4	2×5/12

Table 5.9. Differential transconductor elements.

difference at the outputs. The unusual design is inspired by the multiplier of figure 5.9, which by clever arrangement enables highly linear multiplication with nonlinear devices. Here the same trick is used to realize a simple but highly linear transconductance. Only four transistors instead of eight are needed, with transistor pairs 2 and 4 of figure 5.9 eliminated. A mathematical understanding of this transconductor proceeds easily from the analysis done earlier. Recall the expressions for transistor pairs 1 and 3 from subsection 5.4.1.1:

$$I_1 = \mu_n C_{ox} \left(\frac{W}{L}\right)_{bot} \left((X - V_T)(Y + y - V_{GS-top}) + xy + x(Y - V_{GS-top}) - \frac{(Y - V_{GS-top} + y)^2}{2}\right)$$

and

$$I_3 = \mu_n C_{ox} \left(\frac{W}{L}\right)_{bot} \left((X - V_T)(Y - y - V_{GS-top}) - xy + x(Y - V_{GS-top}) - \frac{(Y - V_{GS-top} - y)^2}{2}\right).$$

For purposes of analyzing figure 5.21, M1 and M3 are sized M_{bot}, M2 and M4 are sized M_{top}, and $x = 0$. The current output of this transconductor is then $I_1 - I_3$:

$$I_{out} = I_1 - I_3 = \left[2\mu_n C_{ox} \left(\frac{W}{L}\right)_{bot} (X - V_T + Y - V_{GS-top})\right] \cdot y.$$

Table 5.9 shows the transistor sizes for the transconductor cell. In the layout, pairs M1-M3 and M2-M4 are each arranged in common-centroid fashion to minimize raw DC offsets.

The schematic for the folding amplifier is shown in figure 5.22. This

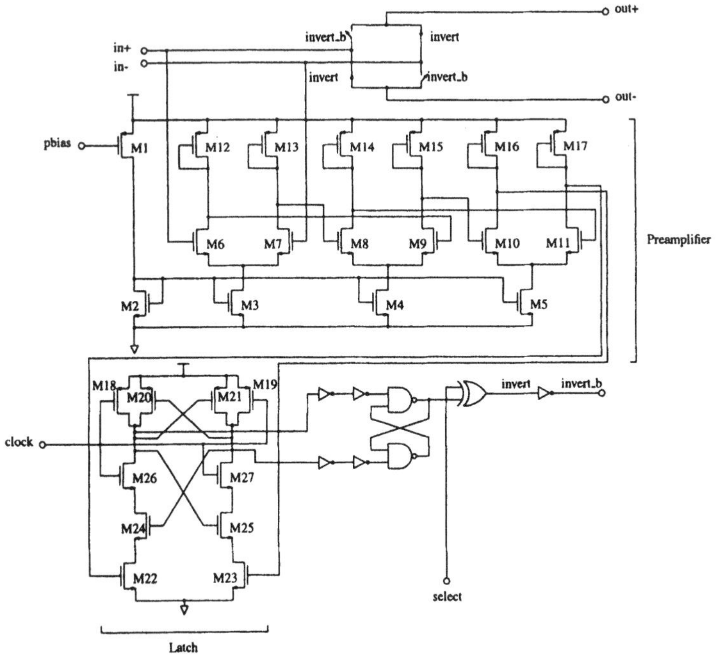

Figure 5.22. Folding amplifier for constant 1-norm controller.

folding amplifier realizes the concept of figure 5.8. A bridge of CMOS switches feeds the input directly to the output, effecting an inversion only if necessary.

Device	Value
M1	49/1
M2-M5	12/1
M6-M7	2×50/10
M8-M11	2/0.24
M12-M13	2/2
M18-M19	1/0.24
M20-M21	3/0.24
M22-M23	2×1/1
M24-M25	2×1/0.24
M26-M27	2/0.24

Table 5.10. Folding amplifier elements.

Necessity is determined in part by the comparator, consisting of a latch [58] and preamplifier, which determines the sign of the input. The switches arrange for the output to be always positive or always negative, depending on the level of the digital *select* input. The frequency and phase of the clock is not critical, and the clock input is simply tied to c_0 from figure 5.17.

5.4.1.5 Forming the matrix rotation operator

As shown in figure 5.5, there are two sources of rotation for the upconverted baseband symbol. One is provided manually, and the other is that demanded by the phase alignment system. The ultimate goal is to perform rotation on a baseband vector, $\vec{b}$, according to

$$R_{\mathrm{T}}\vec{b} = R_{\mathrm{a}}R_{\mathrm{m}}\vec{b} = \begin{bmatrix} \cos\phi_{\mathrm{a}} & -\sin\phi_{\mathrm{a}} \\ \sin\phi_{\mathrm{a}} & \cos\phi_{\mathrm{a}} \end{bmatrix} \begin{bmatrix} \cos\phi_{\mathrm{m}} & -\sin\phi_{\mathrm{m}} \\ \sin\phi_{\mathrm{m}} & \cos\phi_{\mathrm{m}} \end{bmatrix} \vec{b}\,.$$

Performing the matrix multiplication, the matrix operator R_{T} is seen to be

$$R_{\mathrm{T}} = \begin{bmatrix} \cos(\phi_{\mathrm{a}}+\phi_{\mathrm{m}}) & -\sin(\phi_{\mathrm{a}}+\phi_{\mathrm{m}}) \\ \sin(\phi_{\mathrm{a}}+\phi_{\mathrm{m}}) & \cos(\phi_{\mathrm{a}}+\phi_{\mathrm{m}}) \end{bmatrix} \tag{5.5}$$

where the trigonometric identities

$$\cos\phi\cos\theta \mp \sin\phi\sin\theta = \cos(\phi\pm\theta)$$

and

$$\sin\phi\cos\theta \pm \cos\phi\sin\theta = \sin(\phi\pm\theta)$$

have been used to simplify the final result.

The circuit that performs the matrix multiplication that rotates the baseband symbol is described in section 5.5.1.2. Figure 5.23 depicts the computation of the necessary intermediate results,

$$\cos\phi_{\text{out}} = \cos(\phi_{\text{a}} + \phi_{\text{m}}) = \cos\phi_{\text{a}}\cos\phi_{\text{m}} - \sin\phi_{\text{a}}\sin\phi_{\text{m}}$$

and

$$\sin\phi_{\text{out}} = \sin(\phi_{\text{a}} + \phi_{\text{m}}) = \sin\phi_{\text{a}}\cos\phi_{\text{m}} + \cos\phi_{\text{a}}\sin\phi_{\text{m}}.$$

The circuit is a straightforward application of the multiplier cells described

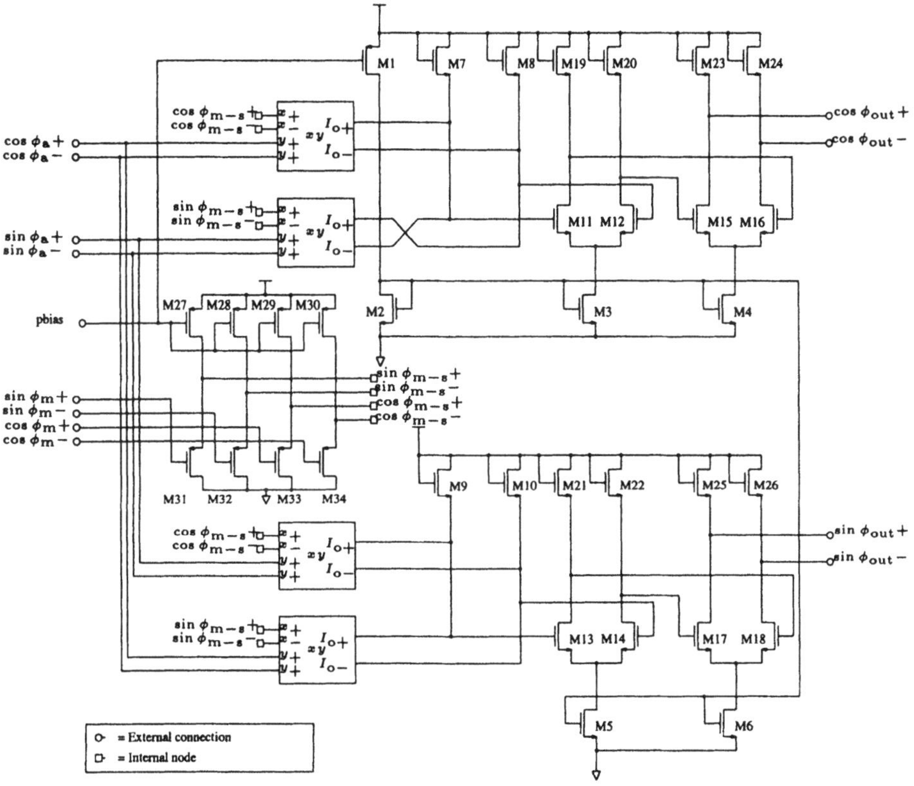

Figure 5.23. Computation of the rotation operator.

in section 5.4.1.1, with the additions and subtractions carried out in the current domain. Resistors M7-M10 accomplish the necessary current-to-voltage conversion, and the subsequent amplifier stages boost the voltages such that full-scale output is approximately 500mV. By design convention, the common-mode level of the differential inputs $\sin\phi_{\text{a}}$, $\cos\phi_{\text{a}}$, $\sin\phi_{\text{m}}$, and $\cos\phi_{\text{m}}$ is maintained at mid-supply (1.25V). The multipliers require a higher common-mode level at x inputs than at the y inputs, so the source followers M31-M34 provide the necessary level shift.

Device	Value
M1	49/1
M2-M6	12/1
M7-M10	6/5
M11-M14	2×6/1
M15-M18	2×8/3
M19-M22	2/8
M23-M26	3.5/10
M27-M30	12/1
M31-M34	2×12/0.5
Quiescent current	826μA

Table 5.11. Rotation operator elements.

5.4.2 Phase alignment system results

An overview diagram of the phase alignment system is shown in figure 5.24. The overall current draw of the phase alignment system is computed as follows:

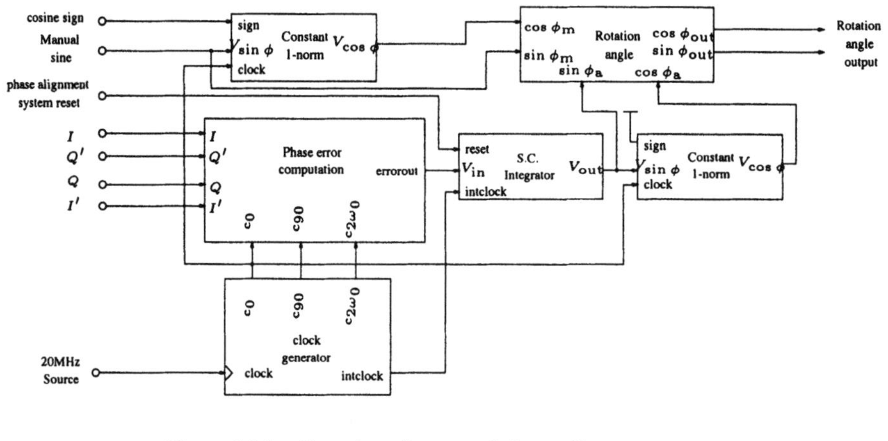

Figure 5.24. Overview diagram of phase alignment system.

the phase error computation block draws 406μA; the clock generator has no static power dissipation; the analog integration requires 524μA; the constant 1-norm controllers account for 2×400μA; and the rotation angle calculation requires 826μA. There is one more contribution to the power dissipation, which is the current required to actually realize the symbol rotation. Because this rotation block is tightly integrated with the upconversion mixer, discussion of its circuit details is deferred to section 5.5. For now, the current draw for rotating the symbol is 962μA, bringing the total current draw of the phase alignment system to 3.5mA. With a 2.5V power supply, the total power dissipation is 8.8mW.

Figure 5.25 summarizes the phase regulation performance of the monolithic prototype. For this test the Cartesian feedback is run in the open-loop configura-

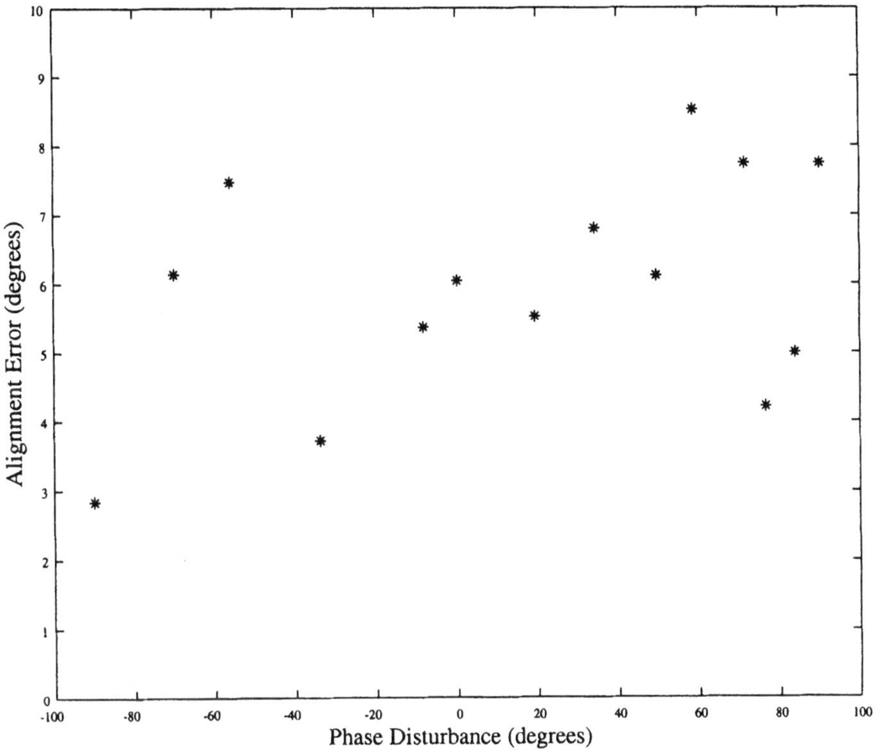

Figure 5.25. Phase alignment performance for a 500mV amplitude, 10 kHz square wave.

tion. The switches in figure 5.3 are configured such that the loop drivers ($H(s)$) are disconnected, while the predistort I/O pins are connected to the upconversion mixer. The test signals, a 10kHz, 300mV amplitude square wave on the I channel and signal ground on the Q channel, are connected to these predistort I/O pins, and the alignment for the 500mV amplitude case, illustrated in figure 5.25, is extrapolated from the resulting data. Static phase misalignments are introduced manually by adjusting a potentiometer, and the phase alignment is switched to active and allowed to settle. Alignment performance is determined by looking at the amplitude of the square waves I' and Q':

$$\Delta\theta = 90 - \arctan\left(\frac{I'}{Q'}\right).$$

The phases of these waveforms relative to the injected square wave, always either 0 or 180 degrees, determine the signs associated with I' and Q'. Figure 5.25

shows that the misalignment never exceeds 9 degrees over the full 180-degree range of disturbances.

The reason for extrapolating data concerns measurement accuracy. The amplitude of Q' decreases rapidly with increasing amplitude of I, making measurement of Q' on the oscilloscope difficult. As discussed in detail in chapter 4, the alignment accuracy for a given output offset in the analog multipliers of figure 5.12 improves with increasing signal amplitude. It is useful to reexamine the equation used to illustrate this fact:

$$Krr' \sin(\Delta\theta) + \delta = 0,$$

where r and r' are the polar magnitudes of the upconverted and downconverted baseband symbols, respectively, K is the constant of proportionality associated with the multipliers, δ is the DC error in computing $IQ' - QI'$, and $\Delta\theta$ is the phase misalignment. This experiment represents the special case for which we can identify $r' \sin(\Delta\theta)$ as, in fact, the amplitude Q', and r as the amplitude I. The constant K is determined by simulation to be approximately 0.306V^{-1}. Substituting and then rearranging terms, we may see the dependence of Q' on the amplitude of the test signal:

$$Q' = r' \sin(\Delta\theta) = \frac{-\delta}{Kr}.$$

For this measurement, a 500mV amplitude for I would have resulted in a Q' of approximately 3mV. For a 300mV amplitude, this figure increases by over 60% to 5mV, which is just barely, but reliably, distinguishable from the noise on the oscilloscope trace. It is important to note that this method breaks down if pursued too aggressively. If Krr' ever has a magnitude less than δ, for instance, no alignment occurs. Once I is reduced to the point where $|\Delta\theta|$ exceeds 45 degrees, the problem simply shifts from measuring Q' to measuring I'.

The procedure for gathering the data of fig 5.25, then, begins with measuring the phase alignment that results from the 300mV signal on the I channel. This phase alignment, $\Delta\theta_{300\text{mV}}$, is used to compute δ:

$$\delta = Krr' \sin(\Delta\theta_{300\text{mV}}).$$

The phase alignment for the 500mV amplitude signal is thus

$$\Delta\theta_{500\text{mV}} = \arcsin\left(\frac{\delta}{0.306\text{V}^{-1} \cdot 500\text{mV} \cdot \left(\frac{500mV}{300mV}\right) \cdot r'_{300\text{mV}}}\right).$$

Figure 5.26 shows the DC error of the phase computation circuit vs. chopping frequency. This error is best described as an *effective* DC offset referred to the input of the first gain stage in figure 5.12. The conditions for this test are

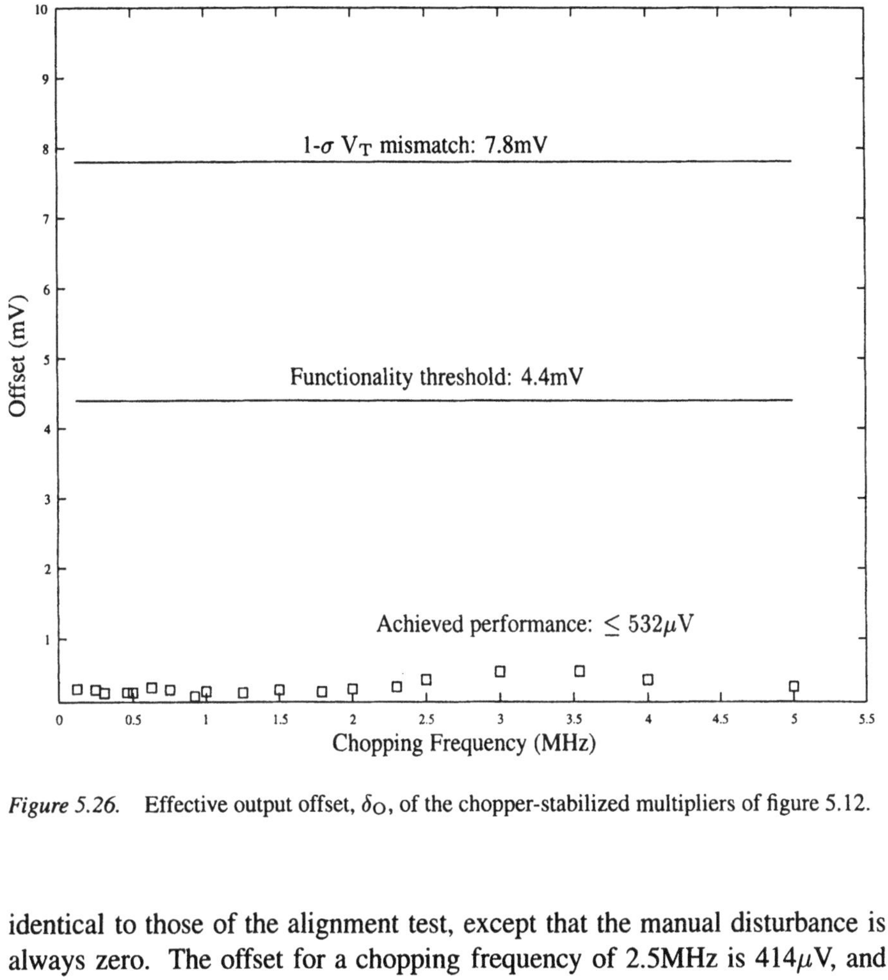

Figure 5.26. Effective output offset, δ_O, of the chopper-stabilized multipliers of figure 5.12.

identical to those of the alignment test, except that the manual disturbance is always zero. The offset for a chopping frequency of 2.5MHz is 414μV, and here the effectiveness of the chopping strategy is evident, as the differential pair of the first gain stage alone has a 1-σ V_T mismatch.[4] of 7.8mV. By way of comparison, complete failure of the the phase alignment system, defined as a 90-degree error for a 500mV amplitude square wave on one channel with the other channel grounded, corresponds to an offset of approximately 4.4mV.

Figure 5.27 is a trace capture of the type of experiment that yields the data of figures 5.25 and 5.26. The Cartesian feedback loop is open, a 500mV amplitude, 10kHz square wave drives the I channel, and the Q channel is grounded.[5] The top two traces show that, initially, the misalignment is manually set to

[4]This mismatch is estimated using the Pelgrom formula [59] Values for the empirical constants in the formula are those suggested by the fabrication facility.

[5]The voltage droop on what is normally the flat part of the square waves is due to the fact that, at the board-level, the predistortion inputs have been AC-coupled.

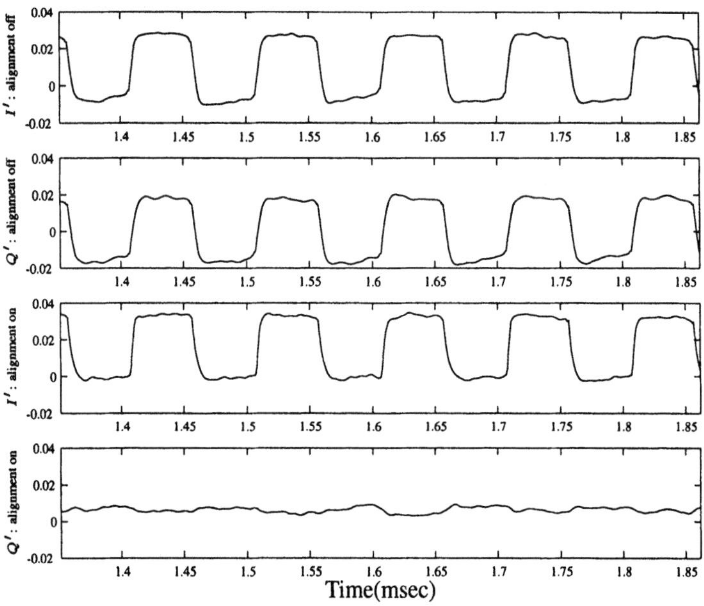

Figure 5.27. Trace capture of a phase alignment experiment. The Cartesian feedback loop is open.

45 degrees. The bottom two traces show the result of turning on the phase alignment system (releasing it from the reset mode described in section 5.4.1.3).

Figure 5.28 serves to illustrate the impact of phase misalignment on the stability margins of the closed-loop CFB system. Dominant-pole compensation is used in the CFB loop, and for the upper two traces the misalignment is manually set to 74 degrees. Overshoot and ringing is evident on these waveforms, and further misalignment causes outright oscillation. For the bottom two traces the phase alignment system is turned on, and one sees the classic first-order step responses that are expected when using dominant-pole compensation.

5.5 The linearization circuitry

For purposes of discussion, "linearization circuitry" now refers to all of the blocks of figure 5.3 outside of the "Phase align" block. Section 5.5.1 describes the circuitry that complete the IC as a fully integrated Cartesian feedback system, and section 5.5.2 details the results of testing.

5.5.1 Circuit details

The list of conceptual blocks needed to complete the CFB system is short: loop driver amplifiers, to provide gain and compensation; up- and downconver-

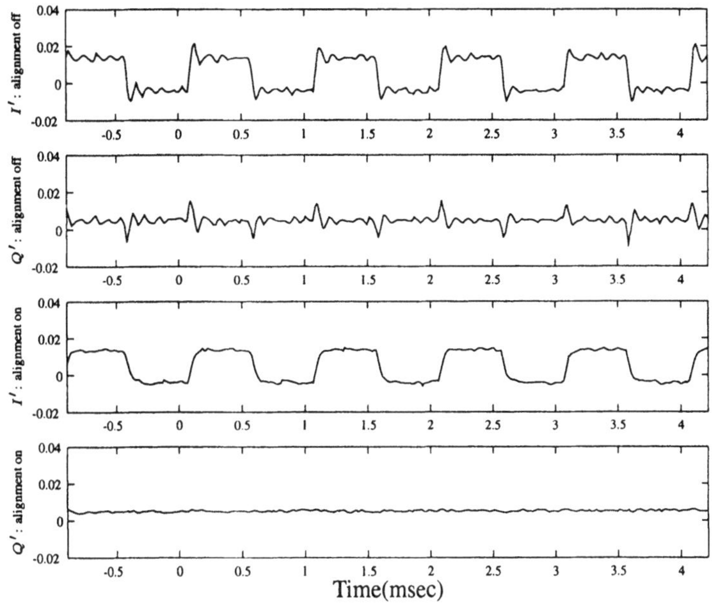

Figure 5.28. Illustration of phase alignment stabilizing the closed-loop CFB system.

sion mixers; and a power amplifier. Realization of these components is now discussed in detail, in addition to the (conceptual) loose ends such as biasing, generating quadrature LO signals, etc..

5.5.1.1 Loop driver

Figure 5.29 depicts the circuit used as the loop driver amplifiers, which map to the $H(s)$ blocks, and the immediately preceding summations, of figure 5.3. Three primary objectives shape the design of the loop drivers. First, their inclusion in the signal path must not disrupt the overall differential architecture. Next, they must have very high DC gain. Last, the user must have some control over the dynamics.

The need to perform a subtraction of two voltage-domain, differential signals complicates the fulfillment of the first objective. Linear subtraction of differential voltages is routinely accomplished using switched-capacitor techniques, as testified to by the vast literature on sigma-delta modulators and pipelined A/D converters. However, for continuous-time subtraction, as needed for this system, two choices are considered. One is to use simple op-amp-and-resistor summers, of the type described in [3]. This idea is rejected on grounds of power consumption. The outputs of the op-amps involved would need to sink or source currents well, something not demanded of op-amps that drive only transistor

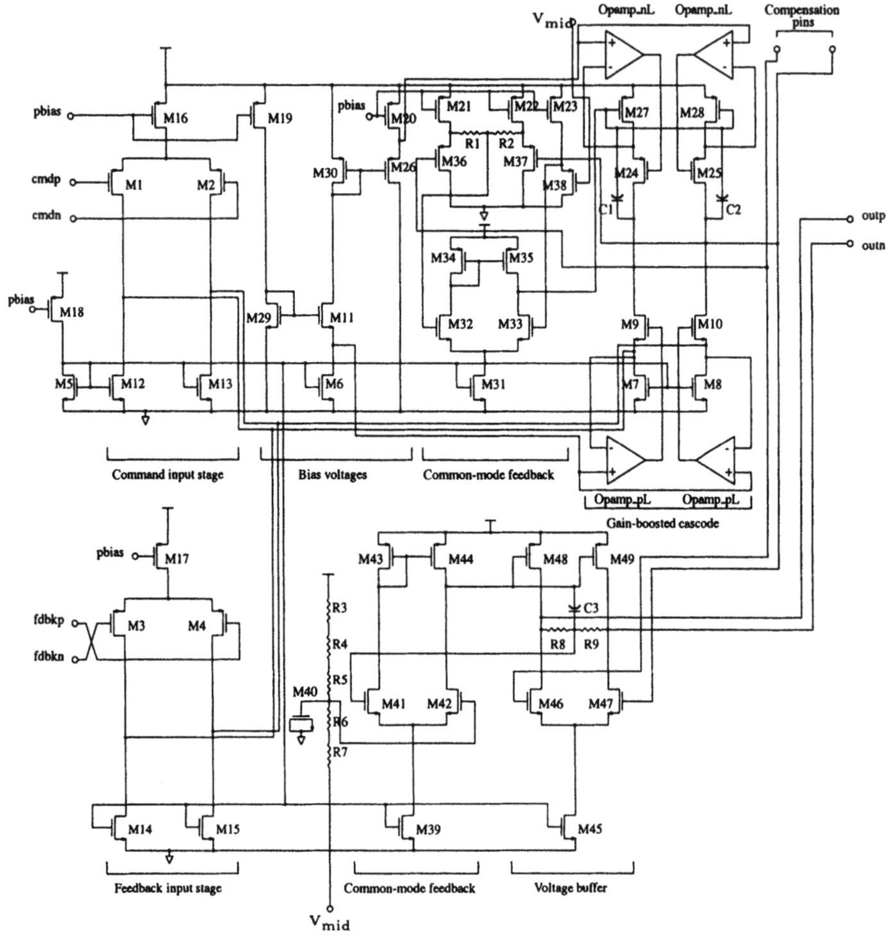

Figure 5.29. Loop driver amplifier.

gates. The other idea is to convert the voltage signals to the current domain, where subtraction is trivial. Figure 5.29 shows that this is the approach actually implemented. Transistors M1-M2 form the differential pair that accepts the "command" input, while transistors M3-M4 accept the "feedback" signal. The current outputs are then subtracted and fed to the gain-boosted cascode. In essence, the whole loop driver structure is simply a fully differential op-amp with an input stage augmented to accommodate a second input.

Differential pairs, of course, have well-known nonlinear transfer characteristics, and in general this would sabotage the linearity of the subtraction just described. The key here is that under closed loop operation, the Cartesian feedback system acts to maintain identical differential currents (of opposite polarity)

Device	Value	Device	Value
M1-M4	2×40/1.5	M36-M38	11/0.24
M5-M11	25/1	M39	12/1
M12-M15	2×25/1	M40	3×15/15
M16-M17	4×49/1	M41-M42	5/10
M18-M26	49/1	M43-M44	10/0.5
M27-M28	2×49/1	M45	32×12/1
M29	5/1	M46-M47	10/0.24
M30	9/1	M48-M49	8×49/1
M31	12.5/1	C1-C2	2.5pF
M32-M33	6/3	R1-R2	60kΩ
M34-M35	20/1	R3-R7	10kΩ
Quiescent current	1.04mA		

Table 5.12. Elements for loop driver amplifier. Quiescent current includes current draw of Opamp_nL and Opamp_pL.

in the two input pairs. To the extent that M1-M4 are matched, the differential voltage inputs are identical. The nonlinearity of each pair is thereby cancelled.

The objectives of high gain and user-controllable loop dynamics are straightforward to satisfy. Gain is assured by the gain-boosting of the folded cascode, and simulations predict a DC gain in excess of 120dB. The idea to make the dynamics user-controllable is inspired by the externally compensated op-amps that were once popular.[6] In these devices, the nodes across which a Miller compensation capacitor would normally be placed are brought out to pins, where the user may choose the value of that capacitor or even implement a more elaborate compensation topology. Strictly speaking, Miller multiplication is not a feature of the folded cascode topology shown in figure 5.29. Nevertheless, the parasitic capacitance always present at the nodes labeled "Compensation pins" does establish a dominant pole. Because use of a network other than a simple capacitance would result in different and potentially interesting loop dynamics, these nodes are brought out to pins.

The output of the loop drivers feed the matrix rotation block of section 5.5.1.2, and it is convenient in turn to feed this output to the inputs of the multipliers that require the higher common-mode voltage. The voltage buffer of the loop driver establishes the common-mode output at 1.75 volts, while also providing a gain of about 5. These properties allow the compensation pins to be kept at the mid-supply common mode voltage while reducing the signal swing. Low signal swing combined with maximum voltage headroom minimizes the impact of nonlinearity at these nodes, allowing the compensation to be set by the user

[6]National's LM301, for instance.

with maximum integrity. The high common-mode for the output eliminates the need for the level-shifting source followers of figure 5.23.

A single-device switch (6/0.24 NMOS) is placed between the compensation pins of each of the loop drivers. During closed-loop operation this switch is open, and has no impact on the operation of the system. During open-loop operation, however, this switch is closed, essentially dropping the gain of the loop drivers to zero. This prevents the loop drivers from being constantly saturated during the open-loop, or "predistort," mode.

5.5.1.2 Analog matrix rotation

Section 5.4.1.5 deals with computing the intermediate products $\sin(\phi_a+\phi_m)$ and $\cos(\phi_a + \phi_m)$ needed for performing analog symbol rotation. Figure 5.30 is a diagram of baseband_g_m, the circuit that takes these products and performs the matrix rotation on the upconverted baseband symbol. Its current outputs are fed directly to the upconversion mixer. Other than perform the needed

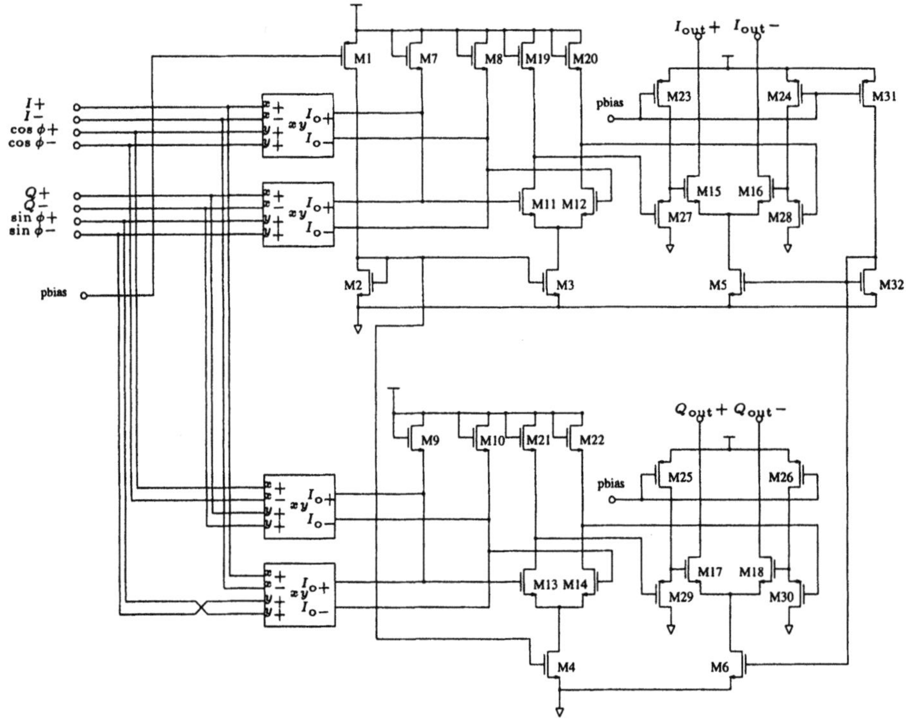

Figure 5.30. Baseband_g_m, a circuit to carry out the matrix rotation.

mathematical operation, the design specifications on this block are few. Good linearity is not a concern: this symbol manipulation occurs in the forward path of the feedback loop, so nonlinearities in baseband_g_m are corrected along with

Device	Value
M1	49/1
M2-M4	12/1
M5-M6	14×60.5/0.5
M7-M10	6/5
M11-M14	2×6/1
M15-M18	4×50.4/0.24
M19-M22	2/8
M23-M26	49/1
M27-M30	20/0.24
M31	6×49/1
M32	60.5/0.5
Output stage quiescent current	4.1mA
Other quiescent current	962μA

Table 5.13. Matrix rotation operator elements.

those of the power amplifier. Also, one expects that the nonlinearities of the power amplifier will be much greater. The exact same reasoning leads to the conclusion that DC offsets in this block are not a concern.

Otherwise, baseband_g_m differs from figure 5.23 primarily in the final stage, where transistors M15-M18 nominally operate with 2mA of bias drain current. Gain of the preceding amplifiers is chosen such that full scale voltage inputs result in saturating these output stages, with 4mA running down one side of the differential pair. The saturating behavior maximizes the power efficiency of the block. The large output currents relax the amplification requirements of the power amplifier.

5.5.1.3 Upconversion mixer

The double-balanced upconversion mixer, shown in figure 5.31, is of the common, Gilbert type [10]. The mixer accepts baseband inputs I and Q as currents from baseband_g_m, and LO voltages derived from an off-chip source (and presumed to be AC coupled). Its output is in the form of a differential current. Here, as with baseband_g_m, linearity is not critical. The only concern is that the devices act as close to the ideal of a perfect switch as possible. This dictates the size of the devices chosen, as well as the drive amplitude from the external LO.

5.5.1.4 Power amplifier

The power amplifier, while conceptually straightforward, has perhaps the greatest potential for disrupting the operation of the chip through unanticipated

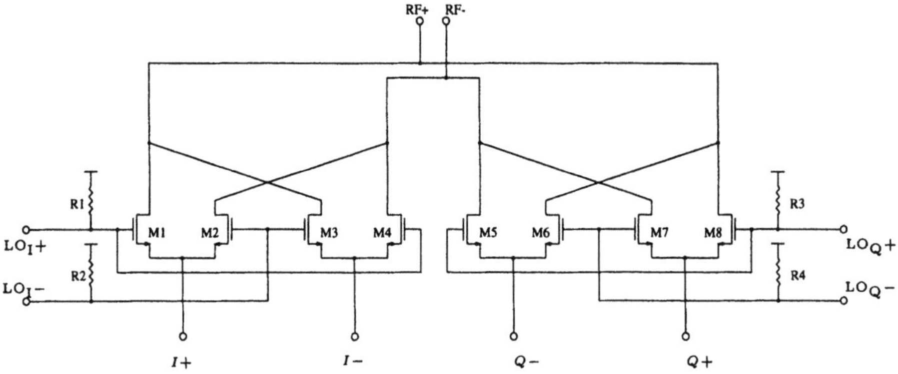

Figure 5.31. Upconversion mixer. All transistors are sized 2× 50.4/0.24, all resistors are 10kΩ.

or poorly modeled parasitic effects. Figure 5.32 is a schematic of the final design.

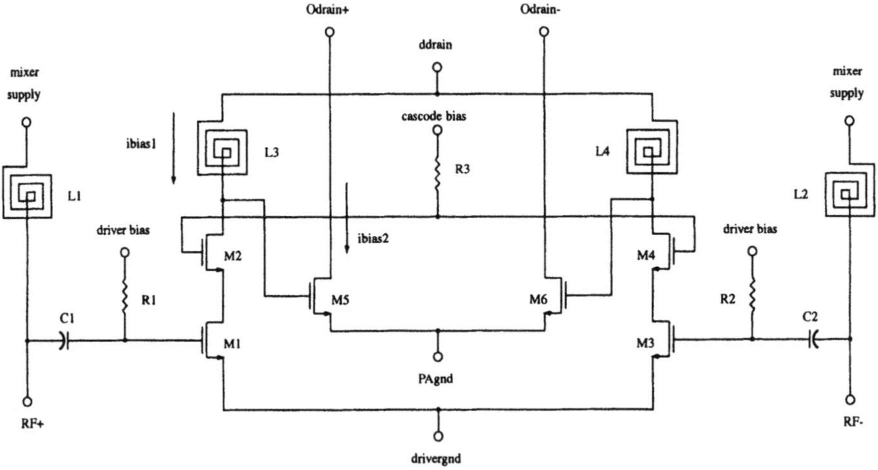

Figure 5.32. Power amplifier.

The primary goals for this circuit are that it provide on the order of 20dBm of output power, and that its large, rapidly varying currents be isolated from the other parts of the IC insofar as is possible. These goals immediately suggest the differential structure shown in figure 5.32. Perfectly differential signals result in no coupling of RF energy into the power supply and ground lines, easing concerns about having a power amplifier on the same die as the linearization system. Moreover, with a 2.5V supply, a fully differential power amplifier

Device	Value
M1-M4	3× 50.4/0.24 , ibias1 = 13mA
M5-M6	5× 50.4/0/24 , ibias2 = 88mA
R1-R2	10kΩ
R3	2kΩ
C1-C2	4pF
L1-L2	5.9nH
L3-L4	4.3nH

Table 5.14. Power amplifier elements.

easily has enough voltage headroom to deliver 20dBm of power to a 50Ω load without the use of an impedance transformation.

Nevertheless, there are many practical pitfalls. Since, for headroom reasons, the differential stages of the PA do not employ tail current sources, this amplifier does *not* possess the desirable common-mode rejection characteristic of many differential amplifiers. Chief among the pitfalls, then, is the possibility of common-mode oscillations. Examining the figure, one sees that the grounds of the two stages, labeled "PAgnd" and "drivergnd," are separated. These do in fact go to separate pins in the package. This is a specific step to reduce the chance of a common-mode oscillation. If one imagines weak coupling between the two grounds, one discovers a common-mode positive feedback loop. Simulations show that sufficient capacitive coupling between these two pins will result in the dreaded oscillatory behavior.

Adjusting for losses in the external balun, this power amplifier ultimately produces 14.2dBm of output power. This power is delivered into a single-ended 50Ω load.

5.5.1.5 Downconverter

The overall linearity of the Cartesian feedback system is limited by the linearity of the downconversion mixer. This is one area in which CFB systems are made more difficult than, say, ordinary op-amp feedback systems. In the latter, the feedback network consists of entirely passive components, which can be relied upon to provide offset-free, extremely linear feedback signals. This emphasis on linearity leads to the choice of the potentiometric downconversion mixer. The mixer, together with the capacitive voltage divider to attenuate the signal from the power amplifier, is shown in figure 5.33.

This realization of the potentiometric mixer differs in two important ways from that described in the paper by Crols and Steyaert [60]. Process variations, in both the nominal g_{ds} of the mixing transistors and in the values of the resistors R1-R2 (made of unsalicided polysilicon), combine to produce large potential variations in the mixer gain. The purpose of the first change is to reduce this

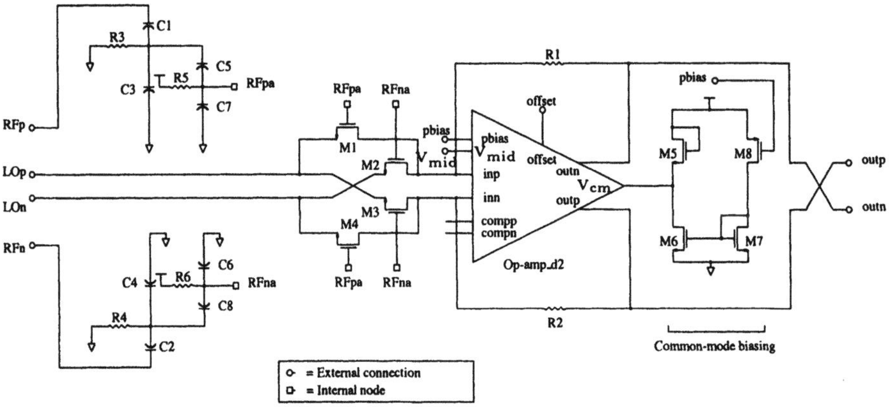

Figure 5.33. Potentiometric downconversion mixer, together with biasing and capacitive RF attenuator.

Device	Value
M1-M4	8×1/1
M5	1/1
M6	12/1
M7	2×12/1
M8	49/1
R1-R2	33.6kΩ
R3-R6	10kΩ
C1-C2	2×10fF
C3-C4	16×10fF
C5-C6	1×10fF
C7-C8	3×10fF
Quiescent current	931μA

Table 5.15. Downconversion mixer elements. Quiescent current includes current draw of Op-amp_d2.

gain variation. As will be discussed in section 5.5.1.7, the source of the "pbias" voltage, until now nearly ubiquitous in the circuit diagrams, is a constant-g_m bias cell. This "pbias" voltage is used in the downconverter to bias M5 such that its g_m is nearly independent of process variations. The LO inputs to the downconverter are AC-coupled, so the nominal gate-to-source voltage of devices M1-M4 is set by the common-mode voltage level at the output of op-amp_d2, which is in turn set by the V_{GS} of M5. The net result is that devices M1-M5 have the same V_{GS}. Now the g_{ds} of devices M1-M4, which are in triode with a drain-to-source voltage of zero, is written

$$g_{\mathrm{ds}} = \mu_{\mathrm{n}} C_{\mathrm{ox}} \left(\frac{W}{L}\right)_{1-4} (V_{\mathrm{GS}} - V_{\mathrm{T}}).$$

The transconductance of M5 is identical in form,

$$g_{\mathrm{m}} = \mu_{\mathrm{n}} C_{\mathrm{ox}} \left(\frac{W}{L}\right)_{5} (V_{\mathrm{GS}} - V_{\mathrm{T}}).$$

Sharing, as these devices do, the same $V_{\mathrm{GS}} - V_{\mathrm{T}}$, it is seen that the g_{ds} of M1-M4 inherits the same immunity to process variations as the g_{m} of M5. The mixer gain variation with process is correspondingly reduced.

The second major departure from [60] is the absence of large capacitors on the inp and inn terminals of op-amp_d2. The purpose of these capacitors is to ensure that these nodes are virtual grounds at RF frequencies, where the op-amp itself is helpless to ensure this condition. These capacitors reduce the bandwidth of the mixer and introduce correspondingly slow poles in the loop transmission of the Cartesian feedback system, so it is desirable to forego these capacitors if possible. The actual voltage amplitude of the RF signal applied to the gates of M1-M4 is small relative to their voltage overdrive $V_{\mathrm{GS}} - V_{\mathrm{T}}$. Accordingly, their g_{ds}'s are never strongly modulated relative to one another. The result is that even with large LO amplitudes, the RF amplitudes at nodes inp and inn never exceed a few millivolts. This sum component of the downconverted RF signal, centered at 4GHz for a 2GHz carrier, is applied to the input differential pair of op-amp_d2. It is then casually and harmlessly dismissed by the op-amp, which is all but insensible to such high-frequency input. In the meantime, the virtual ground at DC frequencies is preserved. Thus the filtering capacitors can be dropped from this particular implementation.

Figure 5.33 shows that the compensation nodes of op-amp_d2 are left unconnected. This works because resistors R1-R2, together with the mixing transistors M1-M4, act to significantly lower the the loop transmission of this op-amp feedback system at all frequencies of interest. The result is that loop crossover occurs early enough in frequency to result in a stable system even in the absence of an explicit compensation network.

A schematic diagram of op-amp_d2 is shown in figure 5.34. It stands out among the several op-amp and op-amp-like structures already discussed in this chapter in that it includes a manual offset trim. Some means of offset removal is nearly always required in direct conversion RF receivers, which is one way to view the feedback path of a CFB system. The need for removing the offset in this case is particularly acute because of the large, DC, closed-loop gain of this circuit block. Using the Pelgrom formula [59], with numerical constant values suggested by the fabrication facility, we may estimate the 3-σ threshold

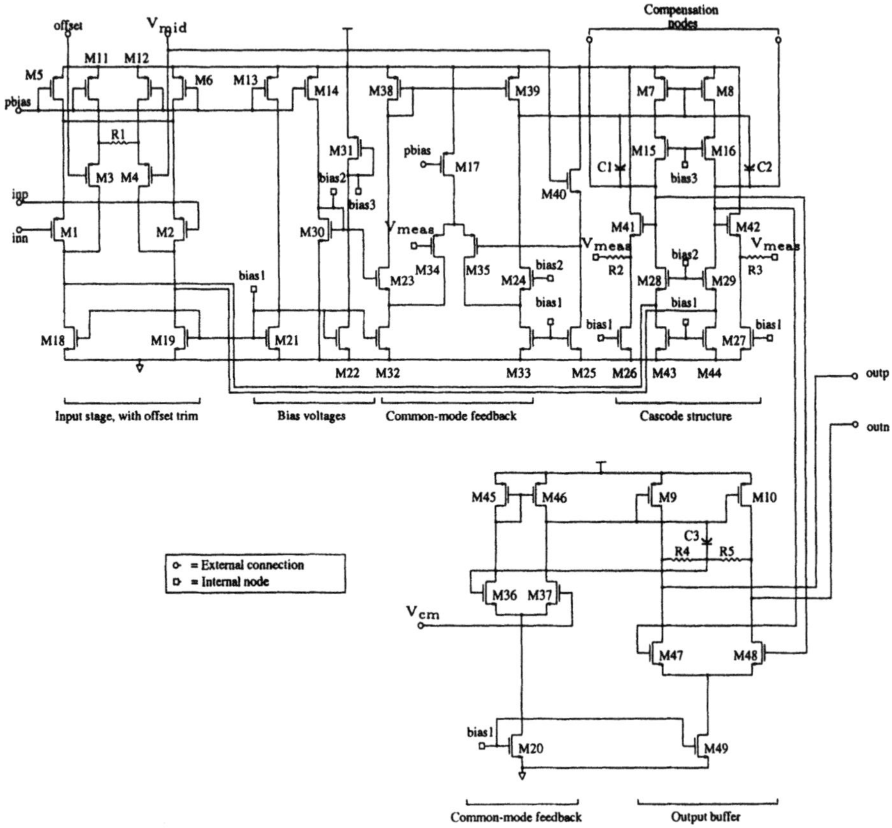

Figure 5.34. Op-amp_d2, a fully differential op-amp for the downconversion mixer.

Device	Value	Device	Value
M1-M2	2×50.4/0.24	M40-M42	12/0.5
M3-M4	25.2/0.24	M43-M44	5×12/1
M5-M10	4×49/1	M45-M46	10/0.5
M11-M17	49/1	M47-M48	10/0.24
M18-M29	12/1	M49	16×12/1
M30	4/2	R1	15kΩ
M31	9/1	R2-R3	30kΩ
M32-M33	2×12/1	R4-R5	2kΩ
M34-M37	5/10	C1-C2	0.3pF
M38-M39	30/0.24	C3	2pF
Quiescent current	860μA		

Table 5.16. Elements for downconversion op-amp.

voltage offset between M1 and M2 to be:

$$3 \times \frac{10\mathrm{mV}}{\sqrt{100.8\mu\mathrm{m} \cdot 0.24\mu\mathrm{m}}} = 6.1\mathrm{mV}.$$

Simulations show the corresponding output offset to be hundreds of millivolts, easily overwhelming the downconverted signal itself. Devices M3-M4 are thus included to allow the re-balancing of the downconverters.

The use of resistive feedback, as shown here, requires that the output of the op-amp be able to sink and source current. The output buffer of figure 5.34 enables this behavior. It also allows for the common-mode output voltage to be set by an input signal, thereby facilitating the aforementioned method of insulating the mixer gain from the process variations.

5.5.1.6 Polyphase filters

The local oscillator source for this IC is external. Some means of deriving quadrature phases from this source is needed, and so the RC sequence asymmetric polyphase filters of figures 5.35 and 5.36 are used for this purpose [61, 62].

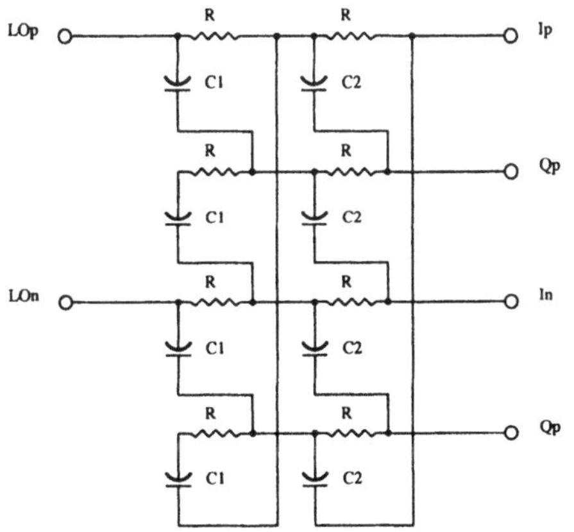

Figure 5.35. A two-stage polyphase filter.

The two-stage filter provides the quadrature phases of the LO for the upconversion mixer, while the three-stage filter is used for the downconversion mixer. This approach is chosen over simply using one filter for both mixers because the two mixers have very different requirements. The downconversion mixer is the most sensitive component in a Cartesian feedback system, and extra care should be taken to ensure that its quadrature LO phases are the highest quality possible. Particularly in the face of component mismatches, the three-stage polyphase filter has the benefit of superior amplitude matching of its two outputs, as well as improved quadrature accuracy compared to a one- or two-stage

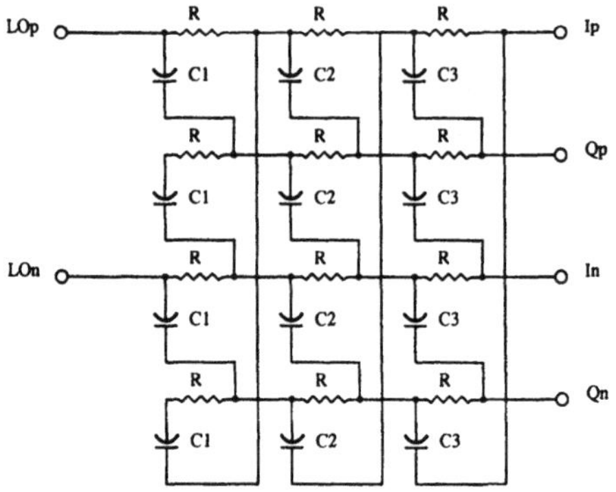

Figure 5.36. A three-stage polyphase filter.

Device	Value
R	160Ω
C1	696fF
C2	355fF
C3	497fF

Table 5.17. Polyphase filter elements.

filter. The upconversion mixer requires high voltage amplitudes delivered to its LO ports, and the two-stage polyphase network results in less attenuation. However, because it is designed as a switching mixer, the upconversion mixer has much less sensitivity to amplitude mismatch than does the downconversion mixer. Because of these different requirements on the part of the two mixers, and in the interest of isolation between the forward and feedback paths, separate polyphase filters are employed.

5.5.1.7 Constant-g_m biasing

Achieving consistent performance in the face of process variations is always a challenge. This is especially true in the case of a large analog system with many interworking functional blocks. Accordingly, a biasing strategy is needed that acts to mitigate performance fluctuations due to process variations for the entire IC.

The strategy is to establish a current, referenced to an external resistor, that corresponds to a constant transconductance for an NMOS device. Figure 5.37 shows a circuit that accomplishes this [10]. The nominal value of the external resistor is 68kΩ. The 'pbias' signal, an input in all of the other diagrams of this chapter, is the output here.

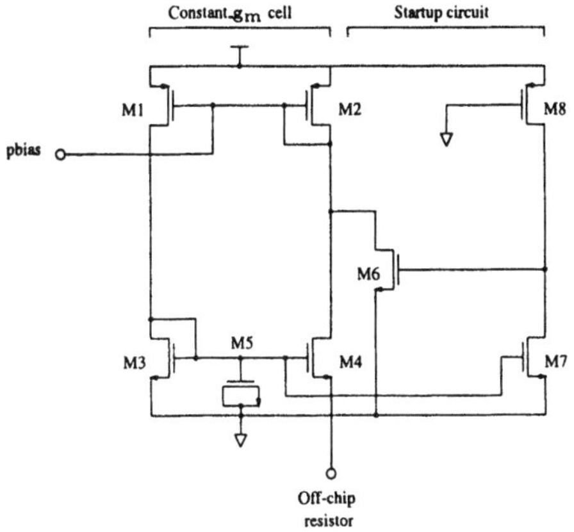

Figure 5.37. Constant_g_m cell, which establishes the voltage 'pbias' for the entire chip.

Device	Value
M1-M2	49/1
M3	1/10
M4	196/0.24
M5	6×200/2
M6	3/0.24
M7	30/0.24
M8	0.62/10
Quiescent current	46.1μA

Table 5.18. Constant-g_m bias cell elements.

This constant-g_m cell has two stable modes of operation. The first is the useful, desirable mode in which it functions as a current reference. In the second, the gates of M3-M5 rest at ground. To avoid this second disastrous mode from occurring, a start-up circuit formed by devices M6-M8 is employed.[7] The gates of M3-M5 being at ground causes M7 to turn off as well. The allows the weak PMOS device M8 to charge up the node formed by its own drain, the drain of M7, and the gate of M6. When this node rises above an NMOS threshold voltage, M6 pulls current through M2 and thereby "jump starts" the constant-g_m cell. The resulting voltage on the gate of M7 turns it on. M7 in turn drags the gate of M6 down to ground, effectively shutting off the start-up circuit.

[7]This circuit idea was given to me by Dr. Hamid Rategh during an informal discussion in the fall of 2000.

5.5.2 Linearization system results

The overall power dissipation of this linearization system is summarized as follows: the loop drivers account for 2×1.04mA of quiescent current; matrix rotation draws 962μA, in addition to 4.1mA for the output stage; the downconverters require 2×931μA; and the constant-g_m cell draws 46.1μA. The 962μA from the matrix rotation is attributed to the phase alignment system of section 5.4.2, and the 4.1mA, together with the current draw of the power amplifier, is not charged to the linearization system as it represents the unlinearized transmitter. The total current draw of the components described in the previous section is thus 4.0mA.

Together with the phase alignment system, we may attribute 7.5mA of current draw to the complete linearization system. All of these power dissipation figures are based on simulation, but they show excellent agreement with what is observed in the laboratory. Some signals routed to pins as test signals use opamp_nL or opamp_pL as buffers, adding 0.56mA to the total predicted current draw. The measured current draw of the chip was 8mA, indicating 20mW of power dissipation. It is important to note that the power consumed in the linearization system is unrelated to the output power of the power amplifier. This system could be applied to a 1W or 1kW output-power PA with no changes that would affect its power dissipation.

A die photo of the complete IC is shown in figure 5.38.

5.5.2.1 Linearization behavior

Taken together, figures 5.39 and 5.40 confirm the functionality of this Cartesian feedback system. The top two traces in each figure are the I and Q baseband waveforms as they enter the upconversion mixer of figure 5.3. They are the *predistorted* baseband symbol. They cancel the nonlinearity and noise in the forward path, and otherwise work to match the outputs of the downconversion mixers as closely as possible to the command inputs. Perfection of the downconversion mixer thus implies perfection of the RF output spectrum. Similarly, noise and/or nonlinearity in the downconverters is referred to the RF output. The bottom two traces are the downconverter outputs.

Figure 5.39 evinces the predistortion action of the Cartesian feedback in a number of ways. For this test, the system is set to perfect phase alignment, the command I channel is driven with a 1kHz sine wave, and the Q channel is grounded.[8] The external compensation pins on the loop drivers are left unconnected. It is seen in this figure that the system performs a gain predistortion,

[8]There are small DC offsets observable in the downconverter output. This is a reflection of the DC offsets in the effective command input. The single-ended-to-differential converters between the board-level command inputs and the inputs of the loop drivers provide one component of these offsets, together with the offsets of the input differential pairs of the loop drivers themselves.

as a 280mVp-p input to the *I* channel of the upconverter is needed to produce a 18mVp-p output on the *I* channel of the downconverter. Inspection with the unaided eye suggests correction of nonlinearity as well, as the sine wave of the downconverter output appears to be of much "higher fidelity" than the uppermost trace. Most remarkable is the dramatic noise reduction that the system accomplishes, as evinced by the relative cleanliness of the bottom two traces compared to the top two.

Figure 5.40 serves as another demonstration of the remarkable predistorting action of the Cartesian feedback system. The experimental conditions remain unchanged from the tests of figure 5.39, except that here the phase misalignment is manually set to 45 degrees. We see that while the output traces of the downconverters are virtually identical to those of the aligned case, the inputs required at the upconverters are very different. In the case of perfect alignment, the system need only provide an appropriately warped sinusoidal input to the *I* channel of the upconverters. Indeed, no sinusoid is evident on the upconverter *Q* channel in figure 5.39. In the case of misalignment this must change, as a sinusoid on the upconverter *I* channel will result in a sinusoid on the downconverter *Q* channel. The solution for the system is to inject a cancelling signal at the *Q* channel of the upconverter, a behavior demonstrated in figure 5.40. Suppression of noise and nonlinearity is also evident here.

Figures 5.39 and 5.40 indicate a significant source of noise somewhere in the forward path of the feedback system.[9] We may eliminate the loop drivers as the source of noise: there is no gain between the error signals and the input-referred disturbance of these blocks, so if the noise were in the loop drivers it would be visible in the downconverter output. This leaves the upconversion mixers (together with the symbol rotation circuitry), the power amplifier, and the downconverters as the possible culprits.

Figure 5.41 is a comparison of the open- and closed-loop spectra of this CFB system. The top spectrum is the result of opening the CFB loop and driving a 1kHz sine wave directly into the *I* channel of the upconversion mixer. The bottom spectrum shows the RF output during closed-loop operation with the phase alignment system active. These figures show that the linearization system produces a third-order harmonic reduction of just under 6dB. However, the noise floor in the closed-loop system is raised by approximately 20dB in relation to that of the open-loop system, providing damning evidence that the downconverter is the dominant source of noise in this system.

How did this happen? Careful reexamination of figures 5.33 and 5.34 immediately places the offset trim under suspicion. Figure 5.34 shows that the offset input is not delivered to the op-amp in a differential way. Rather, V_{mid} is

[9] For this discussion the outputs of the downconverters are temporarily regarded as the system outputs. This, as opposed to the RF output, which is the true output of the system.

the universal mid-supply reference for the entire chip, while the "offset" input is a dedicated, single-ended pin. V_{mid} and the offset input are derived from the external power supply in different ways. It follows that any supply bounce and noise will appear as difference-mode corruptions with respect to this offset trim.

Suspicions increase upon examining the sensitivity of the system to disturbances injected at the offset trim. We start with the gain from the offset input to the output of the op-amp for the stated resistor values, which simulations indicate to be approximately 6 V/V. It is then necessary to refer the output of the downconverter to the input of the upconverter. This is easily done with figure 5.39, which shows that a 280mVp-p sine wave at the upconverter emerges as an 18mVp-p sine wave at the downconverter, representing a gain of 0.06 V/V. Using these two numbers, we arrive at the conclusion that any signal present in the offset trim refers to the input of the upconverters magnified by a factor of approximately 100. Based on figures 5.39 and 5.40, we may roughly estimate the amplitude of disturbances present on this sensitive input to be on the order of a couple hundred microvolts.

In light of this sensitivity of the system to disturbances at the offset trim, various artifacts in the closed-loop output spectrum make sense. The heightened noise floor has already been discussed. The peak centered at the carrier frequency bespeaks DC or near-DC corruptions present in the downconverters. Manual offset adjustments to eliminate this peak are to almost no avail, which makes sense given that millivolt-precision adjustments referred to the upconverter inputs imply a precision of 10μV for manual adjustments here. A more subtle artifact is the heavy presence of the second harmonic in the RF output. This is probably due to the abundance of single-ended-to-differential converters on the board (see appendix B). It is desired to provide balanced baseband inputs to the chip, and active methods are preferred over baluns in order to enable DC probing. Each of these active baluns, however, injects signal at the *second* harmonic into the power supply, and thus corrupts the offset adjustment.

5.5.2.2 Loop stability

As described in section 5.5.1.1, the loop dynamics of the CFB system are externally adjustable. We explore three compensation strategies in this investigation: dominant-pole compensation, uncompensated operation (operation with the compensation pins left floating), and slow-rolloff compensation. The compensation networks are illustrated in figure 5.42. Unless otherwise stated, the LO phases are aligned manually for these experiments.

The performance of the system under dominant-pole compensation is illustrated in figure 5.43. For this experiment a 1kHz square wave is applied to the I command input, and the Q command input is grounded. The 10%-90% rise time is measured to be 39μs, indicating a loop bandwidth of 9kHz. The step response exhibits no overshoot or ringing, confirming the appropriateness of the dominant pole model.

Figure 5.44 depicts the step response of the uncompensated system. The compensation pins are left unconnected (as they were for the traces of figures 5.39 and 5.40). Compared with the dominant-pole case, we see a significant speed increase. The speed increase manifests itself in two important ways. The first is that the noise rejection is vastly improved. This is particularly evident from the Q channel traces of figures 5.43 and 5.44. The second is a substantial shortening of the 10%-90% rise time: 4.3μs, which indicates a bandwidth of 81kHz. This, together with knowing that the compensation capacitor for the dominant-pole case was 1nF, permits estimating the total capacitance, $\mathrm{C_{uc}}$, presented by the bond pads, pins, and board traces connected to the compensation nodes of the loop driver compensation nodes:

$$\frac{1\mathrm{nF} + \mathrm{C_{uc}}}{\mathrm{C_{uc}}} = \frac{81\mathrm{kHz}}{9\mathrm{kHz}}.$$

Solving for $\mathrm{C_{uc}}$ yields an estimate of 125pF.

The performance of the system under slow-rolloff compensation is of particular interest, as it provides experimental confirmation of the idea that a Cartesian feedback system can be stable under 90-degree phase misalignment (see discussion in chapter 4). For purposes of comparison, the slow-rolloff network of figure 5.42 is designed to yield a system with approximately the same loop crossover frequency as in the dominant-pole case.

Figure 5.45 shows a rise time virtually identical to that of the dominant-pole compensated system, measured at 39μs. The noise rejection is again worse than in the uncompensated case, commensurate with the slower loop speed.

Figure 5.46 provides dramatic confirmation of the slow-rolloff ideas of chapter 4. For this experiment, the phase misalignment of the system is manually set to 90 degrees, and the loop compensation is varied. The top two traces show the dominant-pole compensated system under 90-degree misalignment. Substantial overshoot and ringing is visible, indicative of a lightly damped, complex pole pair. Further misalignment to 101 degrees causes outright oscillation. The bottom two traces show the system under slow-rolloff compensation. The step response is remarkably similar to that of a single pole system. In this case, oscillation does not occur until the phase misalignment reaches 117 degrees.

5.6 Summary

This chapter describes a complete Cartesian feedback system integrated on a single die. To the author's knowledge, this is the first time that full integration has ever been achieved. The key to this accomplishment is the discovery of a new, compact solution to the phase alignment problem, the theory of which is described in chapter 4. Also, the use of Cartesian feedback as a way of training a predistorter is proposed as a way of fully exploiting the technique's strengths while circumventing its bandwidth limitation.

Problems in the execution of the downconverter design prevented a clear demonstration of the benefits of Cartesian feedback on the RF output. As discussed in section 5.5.2.1, numerous undesirable artifacts in the RF spectrum can be traced directly to the sensitive offset trims of the downconverter. Two approaches to this problem immediately present themselves. First, bringing the trimming voltages onto the chip differentially would be an unambiguous step in the right direction. Second, the gain from the offset trim to the downconverter output could be substantially reduced. Based on our measurements, we expect these changes would improve considerably the spectrum shown in figure 5.41.

In addition to the overall system, there are a number of smaller results described in this chapter. Section 5.5.1.5 describes two meaningful changes to the traditional potentiometric mixer. The first is the elimination of the large filtering capacitors, and the second is the introduction of a biasing technique to reduce the conversion gain variation over process corners. The chopping strategy described in chapter 4 for overcoming offsets in analog multipliers is successfully implemented here, an idea that is potentially useful in any system requiring analog multipliers. The concept of choosing loop dynamics for robustness to phase misalignment is experimentally demonstrated. Again, to the author's knowledge, this implementation represents the first realization of the Cartesian feedback concept that is stable over a ±90-degree range of phase misalignments.

A complete description of the experimental setup, including the design of the test board, can be found in appendix B.

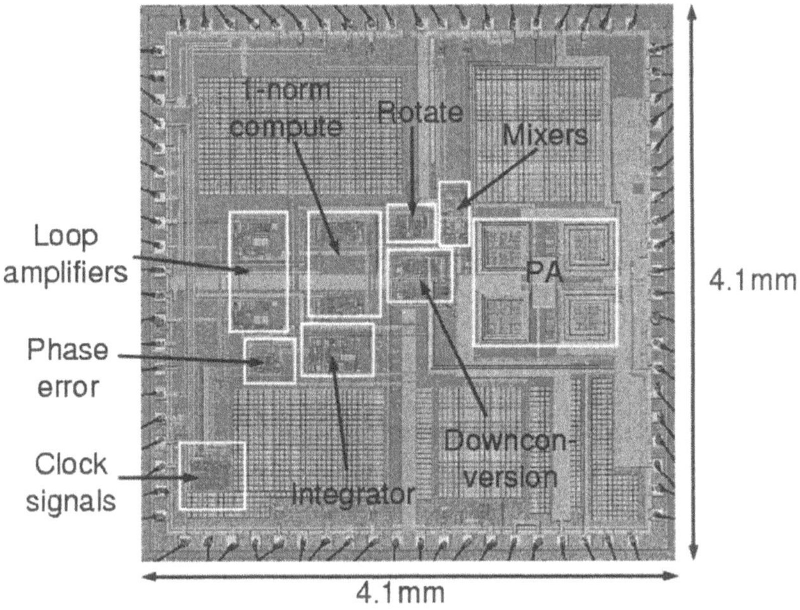

Figure 5.38. Die photo.

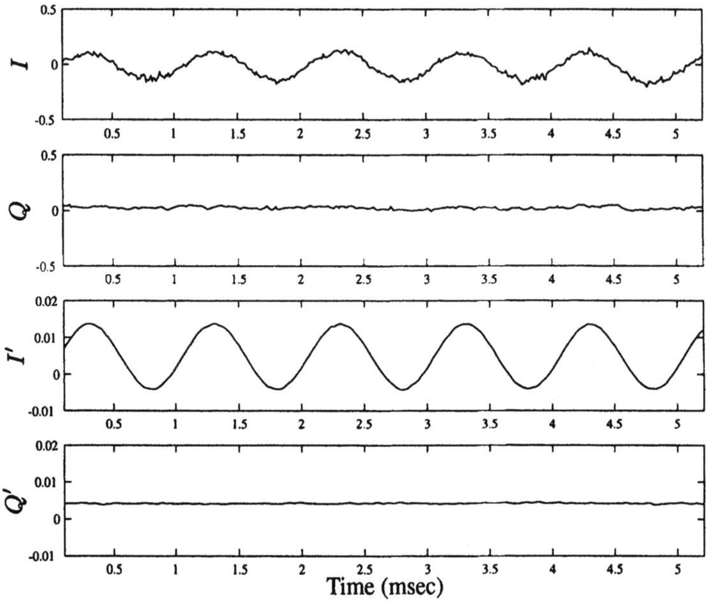

Figure 5.39. Comparison between predistortion inputs and downconverter outputs for no misalignment.

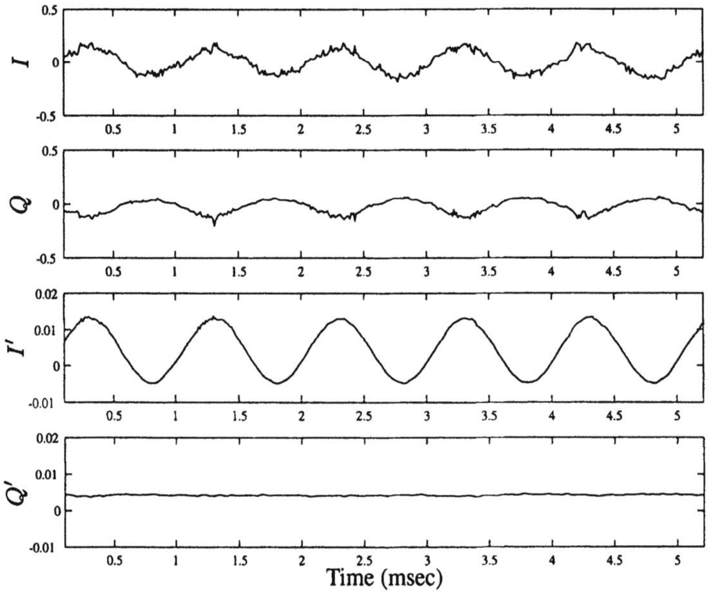

Figure 5.40. Comparison between predistortion inputs and downconverter outputs for 45-degree misalignment.

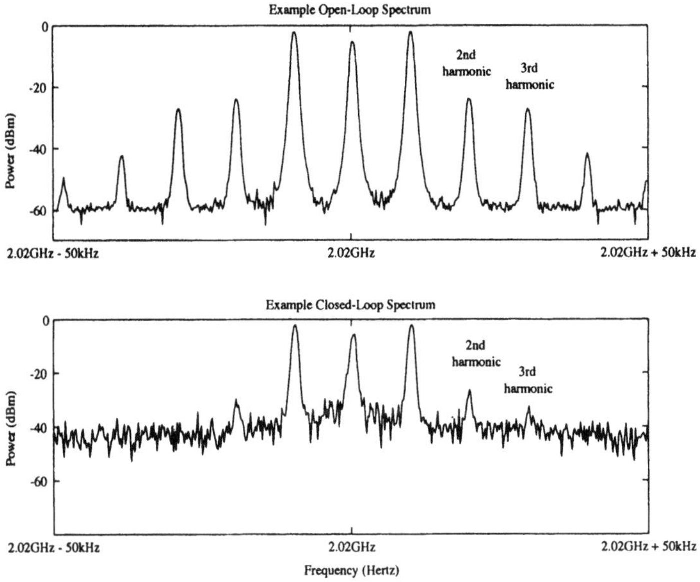

Figure 5.41. Frequency-domain example of linearization behavior.

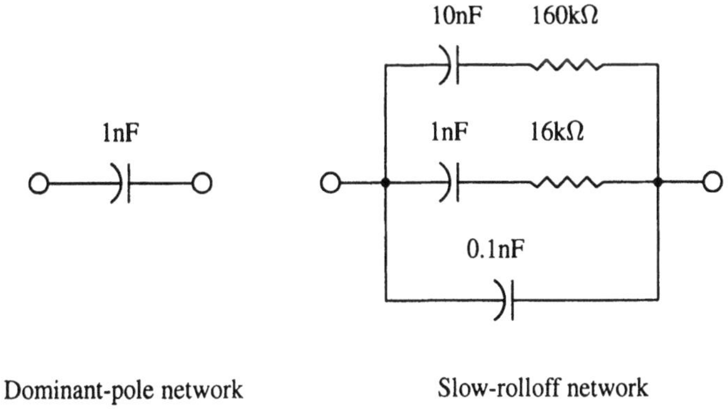

Figure 5.42. Compensation networks used in stability experiments.

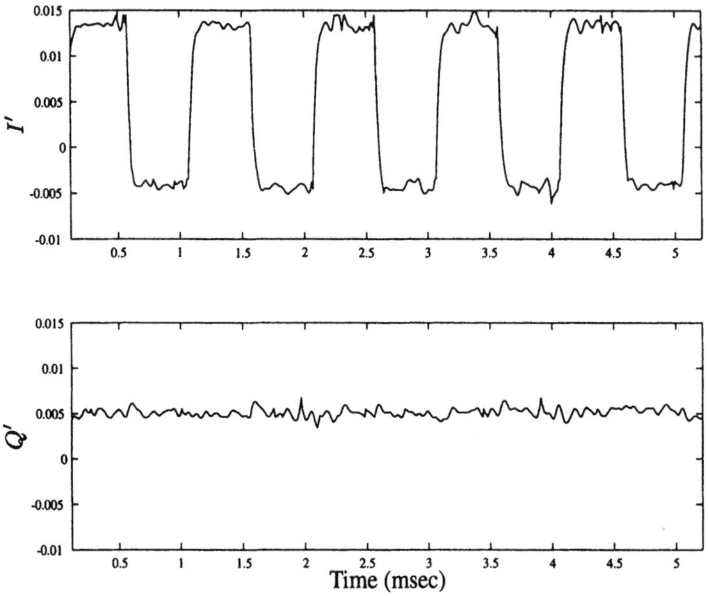

Figure 5.43. Step response of aligned, dominant-pole compensated system.

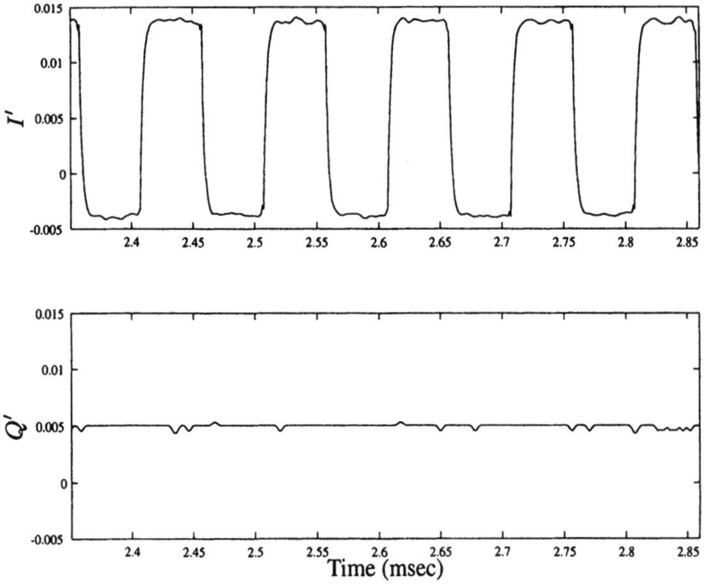

Figure 5.44. Step response of aligned, uncompensated system.

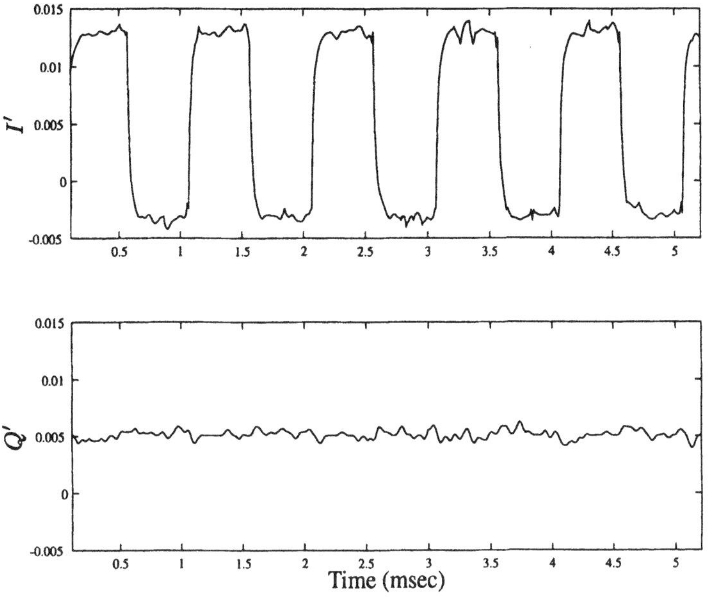

Figure 5.45. Step response of aligned, slow-rolloff compensated system.

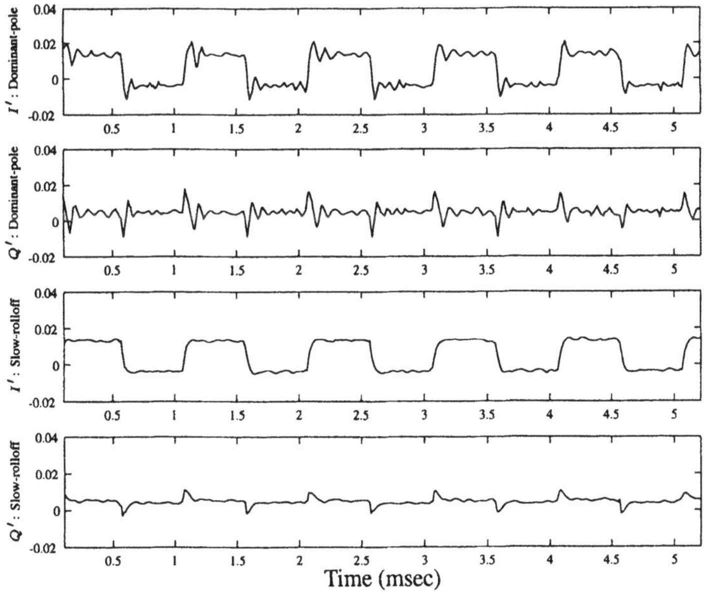

Figure 5.46. Step response comparison between dominant-pole and slow-rolloff compensated systems for 90-degree misalignment.

Chapter 6

CONCLUSION

The purpose of the work described in this book is to contribute to the field of power amplifier linearization. The difficulty of power amplifier modeling is taken as a given, and so as a guiding principle techniques are sought that require little in the way of a PA model. An investigation according to this principle leads naturally to feedback techniques, and two feedback techniques were explored in detail.

This book concludes with a summary of contributions, and thoughts on possibilities for further exploration.

6.1 Summary

Geometric programming has emerged as a powerful optimization technique for analog circuit design. While GP solvers converge quickly and reliably, there is often a great deal of work involved in manipulating a problem into the special form of a geometric program. Chapter 2 demonstrates that the local feedback allocation problem can be solved by geometric programming, and that a variety of amplifier performance metrics can be optimized using this technique. Also shown are optimization curves which illustrate the tradeoffs and different figures of merit.

The first contribution of chapter 4 is to provide a rigorous analysis of the effect of phase misalignment on the stability of a Cartesian feedback system. The slow-rolloff compensation network is proposed to achieve a system that is tolerant to phase misalignments. A new control law is proposed, which leads to an analog, nonlinear phase alignment regulator that is amenable to integration. Finally, as a way of overcoming the problems introduced by DC offsets in the phase alignment system, a new chopper technique for analog multipliers is discovered.

Chapter 5 describes a hardware demonstration of many of the key ideas developed in this book. The first chip to successfully integrate a complete Cartesian feedback system, a phase alignment system, and a power amplifier, is described in detail. The phase alignment and chopper stabilization concepts of chapter 4 are realized successfully in integrated form, and the idea of Cartesian feedback as a way to train a predistorter is proposed. Finally, the feasibility of building a system that is tolerant to large phase misalignments is experimentally demonstrated. Using the slow-rolloff compensation technique, it is demonstrated that even with misalignments of $\pm$ 90 degrees, a Cartesian feedback system can exhibit excellent stability margins.

6.2 Future work

Once the phase alignment problems have been sorted out, Cartesian feedback enables a number of new possibilities. One problem frequently encountered in industry, for instance, is the varying impedance that an antenna can present to the power amplifier. Without careful design, damage to the power amplifier due to unanticipated, excessively high voltages at the output is a possibility. But by enforcing output voltage amplitudes, a stable Cartesian feedback loop quite naturally acts to protect the power amplifier from damaging itself in this case. Also of interest are dynamically biased power amplifiers, where the bias voltages are varied according to the data being transmitted. Despite the boost in efficiency that such systems enjoy, dynamic biasing can effect a time-varying distortion of the RF output. If the system were enclosed in a Cartesian feedback loop, however, this distortion would be suppressed. It would then be possible to be more aggressive with the dynamic biasing algorithm.

Another area for further research concerns the performance limits of a fully integrated Cartesian feedback system. The IC described in this book, for example, does not establish new bounds on achievable linearity, nor does it explore the upper limits of closed-loop bandwidth that are obtainable in a modern CMOS process. The development of an IC to meet cutting-edge performance specifications would be a valuable continuation of this work.

The major idea in this book that goes undeveloped is the use of Cartesian feedback as a way of training a predistorter, and herein lies the most significant avenue for further research. There are a number of intriguing measurement issues associated with filling the predistortion table. How often, for instance, does the table need to be updated? One can imagine either mindless, regular updates, or having the updates be triggered by a sensory event, such as a temperature change. Another issue concerns the details of the table-filling operation itself. For example, an extremely interesting case occurs if the table can be filled very slowly, in the sense of the measurement for each symbol taking a relatively long time. Slow measurements imply long observation times, which can be exploited to minimize the impact of noise in the downconverters on system performance.

Appendix A
The First Prototype of the Phase Alignment Concept

The first realization of the phase alignment concept is a discrete-component prototype [42], designed to work at a carrier frequency of 250MHz. Only the parts of a Cartesian feedback system needed to test phase alignment, namely the upconverter, downconverter, and power amplifier, are assembled.

A.1 Phase shifter

The phase shifter is implemented using a quadrature modulator and an analog control loop, as shown in figure A.1. This control loop forces the sum of the squares of the modulator inputs to

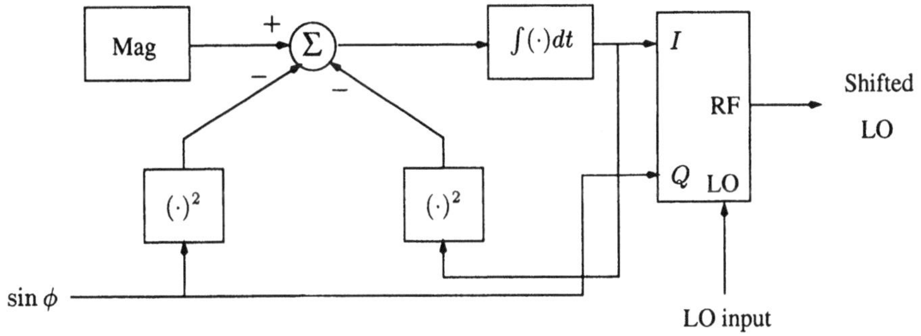

Figure A.1. Phase shifter.

equal a constant ('Mag'), thereby ensuring a constant amplitude for the shifted LO. This actually introduces a small change in the math. We write the shifted LO as

$$I_{\mathrm{LO}} \sin \omega t + Q_{\mathrm{LO}} \cos \omega t.$$

To within a multiplicative constant, this is equal to

$$\cos \phi \sin \omega t + \sin \phi \cos \omega t.$$

For this prototype, then, the analog input is proportional to the *sine* of the phase shift. A functional block labeled 'arcsin(·)' effectively exists between the integrator output and phase shifter input.

The implemented phase shifter is shown in figure A.2. Because of the squaring functional

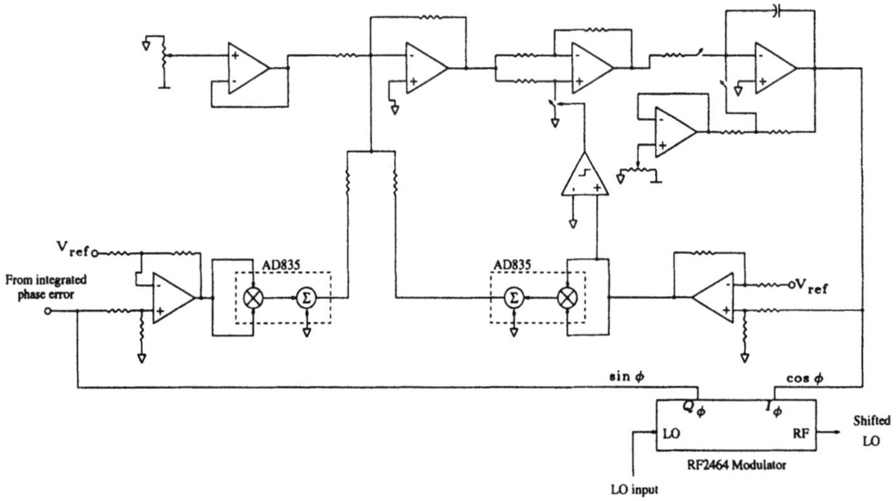

Figure A.2. Phase shifter implementation.

blocks, there are in general two values of cos ϕ that would satisfy the control loop. The sign of the incremental gain around the loop is positive for one solution and negative for the other. The comparator in figure A.2 ensures stability by switching the sign of the loop gain based on the current value of cos ϕ.

The switches on all of the integrators are purely for testing purposes and are manually operated. These familiar "3-mode integrators" allow the outputs of the integrators to be held at their last value, to be manually adjusted with potentiometers, or to operate normally as integrators.

A.2 Phase error and integrator

The phase alignment system core, shown in figure A.3, represents a straightforward mapping from the concept of chapter 4 to op-amp building blocks. For reasons discussed in that chapter, it is necessary to trim the output offsets of the AD835 multipliers (from Analog Devices). This trimming is crudely accomplished via a potentiometer connected to the summing input of one of the multipliers.

A.3 Test results

The prototype shown in figures A.2 and A.3 is built and tested in a 250 MHz RF system. Figure A.4 shows the test setup. Off-the-shelf discrete components are used for the mixers, and the amplifier is the HP8347 bench-top, leveled-power amplifier. Not shown is a resistive voltage divider at the input of the downconversion mixer. An additional phase shifter is inserted in the control path and is manually controlled. By varying this phase shift, we simulate drift normally due to temperature and aging. Figure A.5 shows the outcome of this experiment. The I channel is driven with a 50mV sinusoid, and the Q channel is grounded. It can be seen that

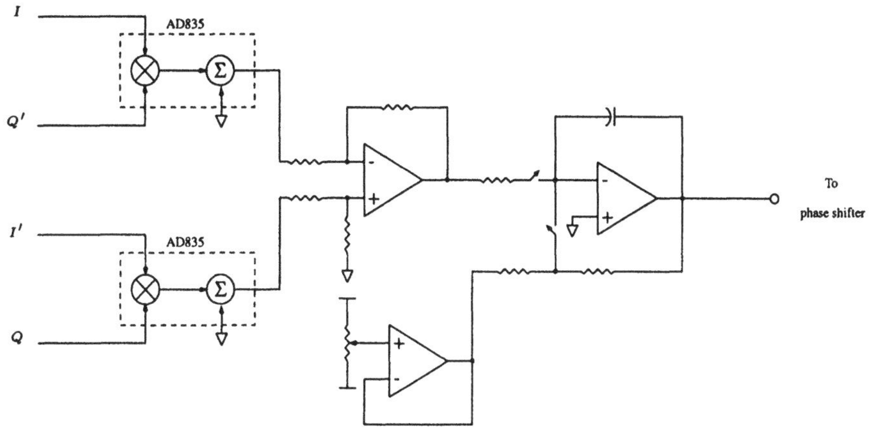

Figure A.3. Phase error computation and integration.

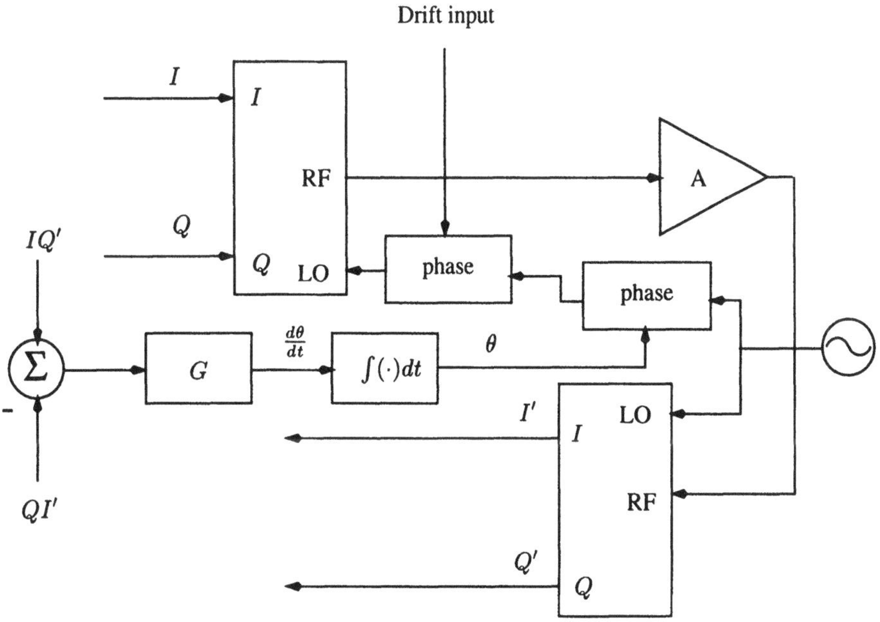

Figure A.4. Test setup.

our prototype automatically and continuously compensates for misalignments as large as ±88 degrees. Alignment to within 3.8 degrees is maintained over this entire range of disturbances.

Figure A.6 shows system performance as the frequency of the input sinusoid is varied. It is seen that performance deteriorates rapidly above 2 MHz. This is due to the op-amp used to build the subtractor (National Semiconductor's LMC6484), which has a gain-bandwidth product of 1.5MHz.

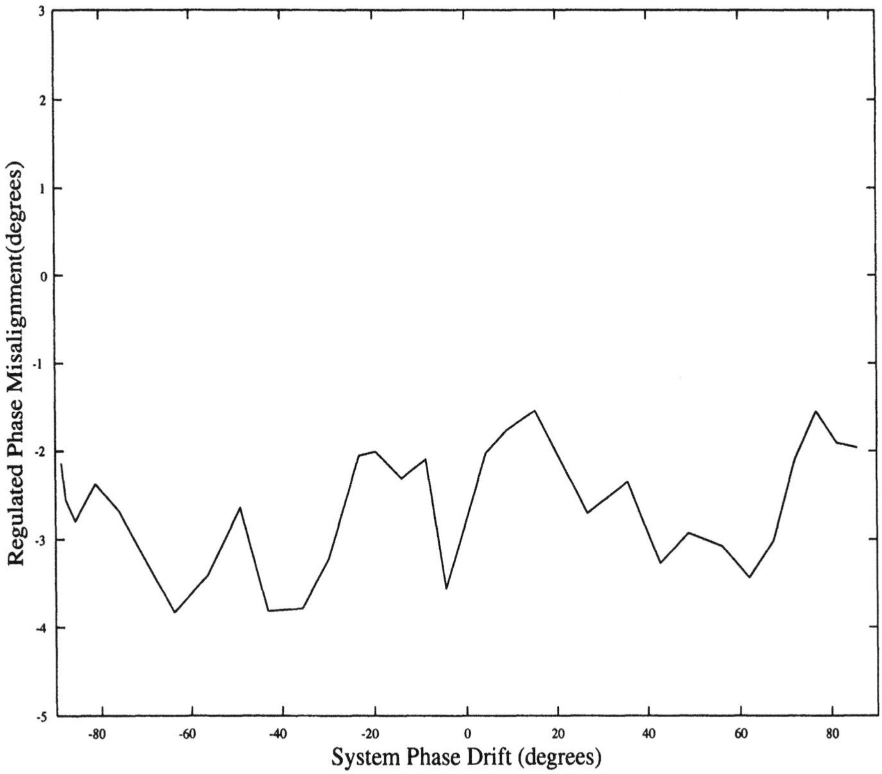

Figure A.5. Measured phase alignment vs. system drift

The following table [1] compares the prototype with other examples from the literature.

Work	Accuracy	Baseband Bandwidth	Carrier (GHz)
This work	**3.8 degrees**	**2 MHz**	**.25**
[36]	10 degrees	21 kbaud	.90
[30]	not reported	25 kHz	.22
[18]	15 degrees	500 kbaud	.90

Table A.1. Comparison with examples from the literature.

[1] Accuracy value for Ohishi et al. is inferred.

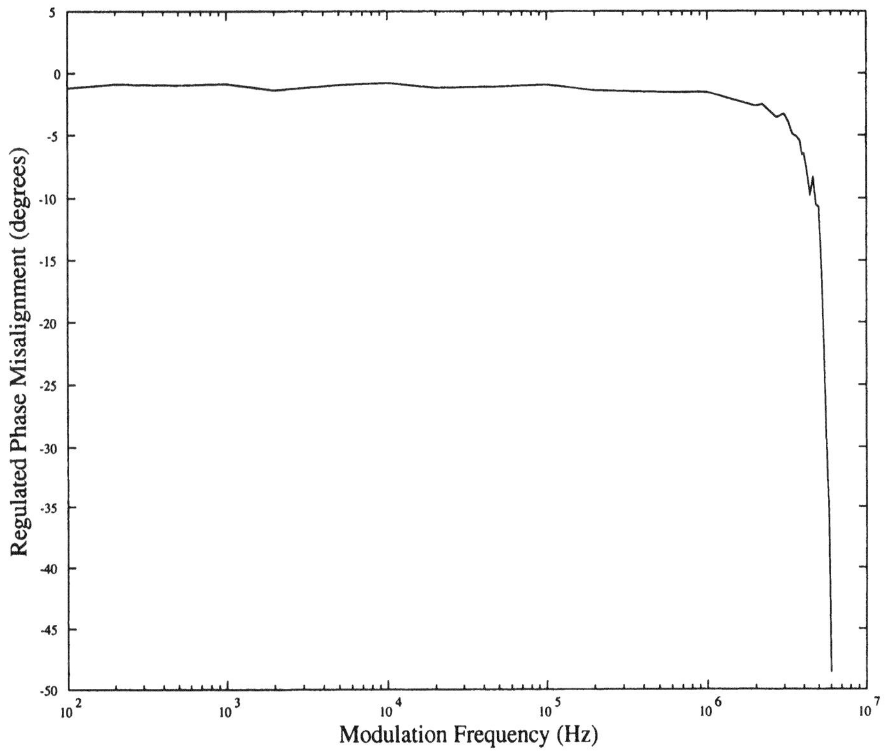

Figure A.6. Phase alignment vs. baseband frequency

Appendix B
The Experimental Setup for CFB IC

This appendix describes the test board for the fully integrated prototype. The final board is fairly large, measuring approximately 8x9 inches. It has three layers of interconnect. The upper and lower layers are used for ordinary traces, while the middle layer is a ground plane.

B.1 Single-ended-to-differential conversion

All of the critical analog signals on the IC are fully differential. This complicates the testing somewhat, as test instruments in the laboratory normally provide (and accept) single-ended signals. It is therefore necessary to convert between single-ended and differential signals at the board level.

Further complicating the design is that it is desirable to use DC and low-frequency input signals, making it impossible to use an ordinary transformer. Figure B.1 shows the design solution for single-ended-to-differential conversion used on the test board (labeled "S/D" in figure B.3). Nominally, the common-mode level for baseband differential signals is 1.25V. Trimming to achieve this level is done by varying the resistor indicated in figure B.1. The predistortion inputs on the chip actually require an S/D converter with two special characteristics: the common-mode level needed is 1.75V, and the output resistance of the converter needs to match that of the on-chip driver used for closed-loop operation. The resistor values for this S/D converter are chosen to satisfy these two constraints.

Differential-to-single-ended conversion is a simpler problem, involving two op-amps as voltage buffers and one op-amp to perform a subtraction. These D/S converters nominally drive the capacitive inputs of an oscilloscope. Here the choice of the LM6134 op-amp proves challenging, as it is designed with unusually low stability margins. Efforts at compensation aside, it is ultimately necessary to configure the oscilloscope such that it presents a resistive, 50Ω load to the D/S output.

B.2 Clock reference

Figure B.2 shows the crystal oscillator circuit used to provide a stable clock reference. Complicating this normally straightforward design is that the only crystal oscillators available currently are for 5V or 3.3V supplies, while the IC operates off a 2.5V supply. Not wishing to add yet another supply line to the board, we make do with the ±2.5V supply already available.

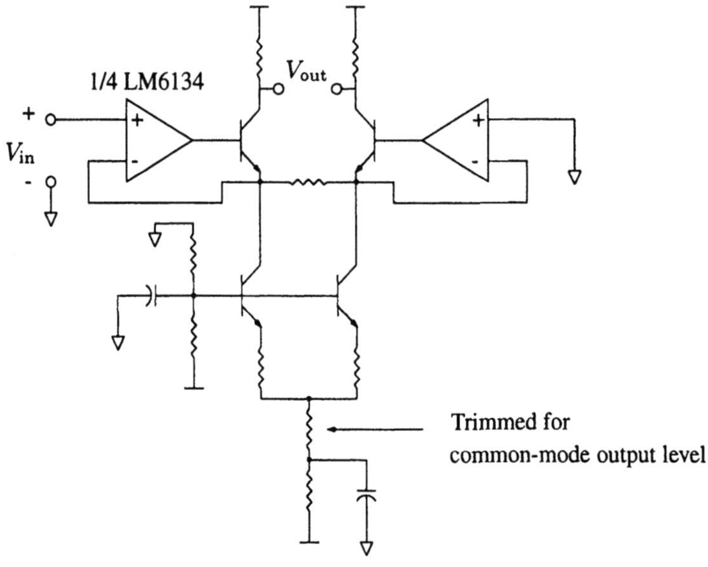

Figure B.1. Converting from single-ended to differential signals.

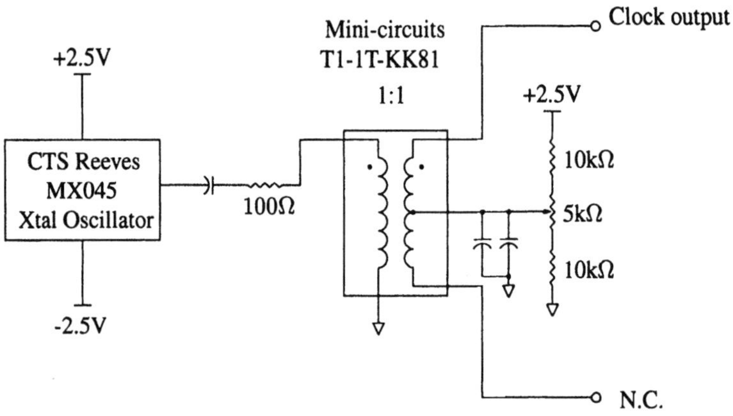

Figure B.2. The on-board clock reference.

The transformer in figure B.2 exploits the fact that if the oscillator output has a peak-to-peak amplitude of 5V, each of the balanced outputs on the secondary side will have peak-to-peak amplitudes of 2.5V. The center tap of the secondary winding is biased by the potentiometer such that the actual clock output of this circuit swings between ground and 2.5V.

B.3 Overview of test board

A simplified diagram of the test board is shown in figure B.3. The "monitor" signals, *Imonitor* and *Qmonitor*, are used to monitor the baseband inputs to the upconverter mixers. These are the "predistortion inputs" of figures 5.39 and 5.40, for example. The nominal values for the voltage

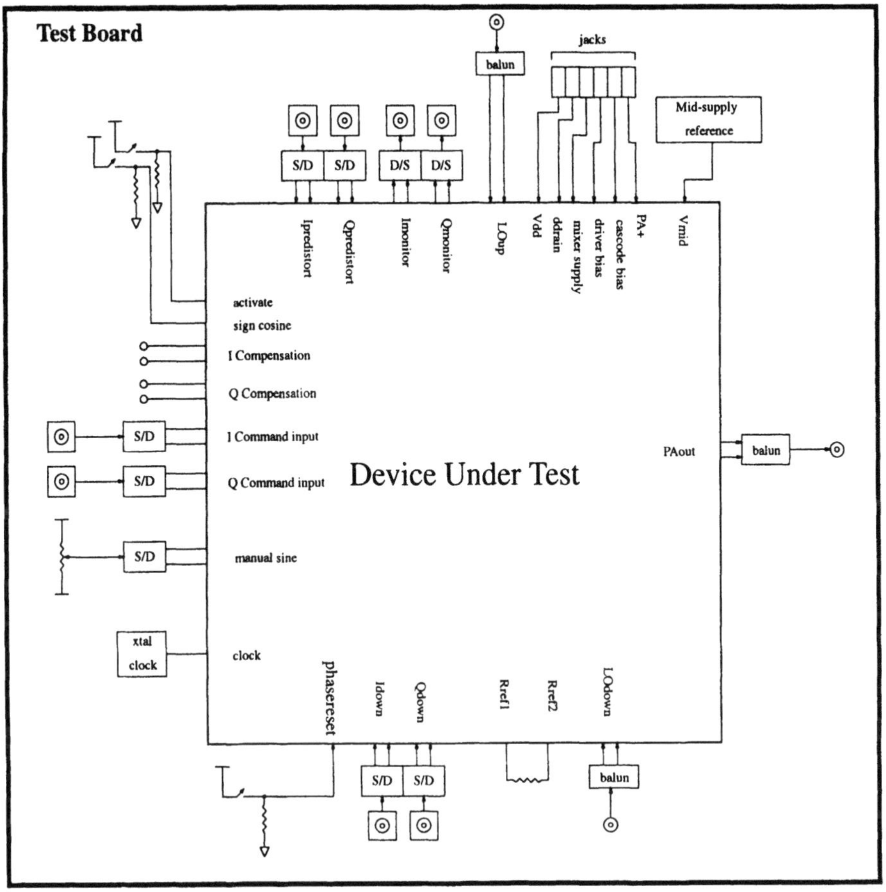

Figure B.3. The test board.

supplies to the board are as follows: Vdd is 2.5, Ddrain and cascode bias are 2.1V, driver bias is 1.1V, the (upconversion) mixer supply is 3.5V, and the PA drains are biased at 2.0V. As noted in section B.2, a -2.5V supply is also available for various board-level components.

References

[1] H. S. Black, "Stabilized feedback amplifiers," *Bell System Technical Journal*, vol. 13, pp. 1–18, 1934.

[2] H. W. Bode, *Network Analysis and Feedback Amplifier Design*, Van Nostrand, New York, 1945.

[3] J. K. Roberge, *Operational Amplifiers: Theory and Practice*, John Wiley and Sons, Inc., New York, New York, 1975.

[4] J.L. Dawson, S. Boyd, M. Hershenson, and T.H. Lee, "Optimal allocation of local feedback in multistage amplfiers via geometric programming," *IEEE Transactions on Circuits and Systems–I: Fundamental Theory and Applications*, vol. 48, pp. 1–11, January 2001.

[5] S. Boyd and L. Vandenberghe, "Introduction to convex optimization with engineering applications," Course Notes, 1997, http://www.stanford.edu/class/ee364/.

[6] J. Ecker, "Geometric programming: methods, computations and applications," *SIAM Rev.*, vol. 22, no. 3, pp. 338–362, 1980.

[7] Y. Nesterov and A. Nemirovsky, *Interior-point polynomial methods in convex programming*, vol. 13 of *Studies in Applied Mathematics*, SIAM, Philadelphia, PA, 1994.

[8] P. R. Gray and R. G. Meyer, *Analysis and Design of Analog Integrated Circuits*, John Wiley and Sons, New York, NY, 1993.

[9] W. M. Siebert, *Circuits, Signals, and Systems*, The MIT Press, Cambridge, MA, 1993.

[10] T. H. Lee, *The Design of CMOS Radio-Frequency Integrated Circuits*, Cambridge University Press, New York, NY, 1998.

[11] R. J. Duffin, E. L. Peterson, and C. Zener, *Geometric Programming — Theory and Applications*, Wiley, 1967.

[12] M. Hershenson, S. Boyd, and T. H. Lee, "GPCAD: A tool for CMOS op-amp synthesis," in *IEEE/ACM International Conference on Computer Aided Design*, San Jose, CA, 1998, pp. 296–303.

[13] M. Hershenson, S. P. Boyd, and T. H. Lee, "Automated design of folded-cascode op-amps with sensitivity analysis," in *International Conference on Electronics, Circuits and Systems*, September 1998, vol. 1, pp. 1–1.

[14] M. Hershenson, S. Mohan, S. Boyd, and T. H. Lee, "Optimization of inductor circuits via geometric programming," in 36th *Design Automation Conference*, 1999, pp. 994–998.

[15] K. O. Kortanek, X. Xu, and Y. Ye, "An infeasible interior-point algorithm for solving primal and dual geometric progams," *Mathematical Programming*, vol. 76, no. 1, pp. 155–181, 1997.

[16] M. Hershenson, S. Boyd, and T. H. Lee, "Optimal design of a CMOS op-amp via geometric programming," in *Applied and Computational Control, Signals, and Circuits*, B. Datta, Ed., vol. 2. Birkhauser, 2000.

[17] W. J. Rugh, *Nonlinear System Theory: The Volterra/Wiener Approach*, The Johns Hopkins University Press, Baltimore, Maryland, 1981.

[18] M. Johansson, "Linearization of RF power amplifiers using Cartesian feedback," Tech. Rep., Lund University, 1991, Thesis for the degree of Teknisk Licentiat.

[19] R. C. Tupynamba and E. Camargo, "MESFET nonlinearities applied to predistortion linearizer design," in *IEEE MTT-S Digest*, 1992, pp. 955–958.

[20] N. Imai, T. Nojima, and T. Murase, "Novel linearizer using balanced circulators and its application to multilevel digital radio systems," *IEEE Transactions Microwave Theory and Techniques*, vol. 37, pp. 1237–1243, 1989.

[21] G. Karam and H. Sari, "Implementation and performance of data predistortion with memory in digital microwave radio," in *Proceeding of the IEEE Global Telecommunications Conference (GLOBECOM)*, 1989, vol. 1, pp. 400–405.

[22] H.W. Kang, Y.S. Cho, and D.H. Youn, "On compensating nonlinear distortions of an OFDM system using an efficient adaptive predistorter," *IEEE Transactions on Communications*, vol. 47, pp. 522–526, 1999.

[23] B. Razavi, *RF Microelectronics*, Prentice-Hall, Inc., Upper Saddle River, NJ, 1998.

[24] A.A.M. Saleh and D. Cox, "Improving the power-added efficiency of FET amplifiers operating with varying-envelope signals," *IEEE Transactions on Microwave Theory and Techniques*, vol. 83, pp. 51–56, 1983.

[25] D. Su and W. McFarland, "An IC for linearizing RF power amplifiers using envelope elimination and restoration," in *ISSCC Digest of Technical Papers*, 1998, pp. 54–55.

[26] P. Nagle, P. Burton, E. Heaney, and F. McGrath, "A wide-band linear amplitude modulator for polar transmitters based on the concept of interleaving delta modulation," *IEEE Journal of Solid-State Circuits*, vol. 37, pp. 1748–1756, 2002.

[27] D.C. Cox, "Linear amplification with nonlinear devices," U.S. Patent # 3,777,275, December 4, 1973.

[28] D.C. Cox, "Linear amplification with nonlinear components," *IEEE Transactions on Communications*, vol. COM-22, pp. 1942–1945, 1974.

[29] D.C. Cox and R.P. Leck, "A VHF implementation of a LINC amplifier," *IEEE Transactions on Communications*, vol. COM-24, pp. 1018–1022, 1976.

[30] Majid Boloorian and Joseph McGeehan, "The frequency-hopped Cartesian feedback linear transmitter," *IEEE Trans. on Vehicular Technology*, vol. 45, pp. 688–706, 1996.

[31] M. Johansson and L. Sundström, "Linearisation of RF multicarrier amplifiers using Cartesian feedback," *Electronics Letters*, vol. 30, pp. 1110–1111, 1994.

[32] A. Bateman, D. Haines, and R. Wilkinson, "Linear transceiver architectures," in *IEEE Conference on Vehicular Technology*, 1988, pp. 478–484.

[33] M. Johansson and L. Sundström, "Linearised high-efficiency power amplifier for PCN," *Electronics Letters*, vol. 27, pp. 762–764, 1991.

[34] M. Briffa and M. Faulkner, "Gain and phase margins of Cartesian feedback RF amplifier linearisation," *Journal of Electrical and Electronics Engineering, Australia*, vol. 14, pp. 283–289, 1994.

[35] M. Briffa and M. Faulkner, "Dynamically biased Cartesian feedback linearization," in *Proceedings of the 1993 IEEE Vehicular Technology Conference*, 1993, pp. 672–675.

[36] Y. Ohishi, M. Minowa, E. Fukuda, and T. Takano, "Cartesian feedback amplifier with soft landing," in *3rd IEEE International Symposium on Personal, Indoor, and Wirless Communications*, 1992, pp. 402–406.

[37] M. Faulkner, D. Contos, and M. Briffa, "Performance of automatic phase adjustment using supply current minimization in a RF feedback lineariser," in *The 8th IEEE International Symposium on Personal, Indoor, and Mobil Radio Communications*, 1997, pp. 858–862.

[38] D. Cox, "Linear amplification by sampling techniques: A new application for delta coders," *IEEE Transactions on Communications*, vol. COM-23, pp. 793–798, 1975.

[39] G. Han and E. Sánchez-Sinencio, "CMOS transconductance multipliers: A tutorial," *IEEE Transactions on Circuits and Systems-II*, vol. 45, pp. 1550–1563, 1998.

[40] A. Bakker, K. Thiele, and J. Huijsing, "A CMOS nested chopper instrumentation amplifier with 100nV offset," in *ISSCC Digest of Technical Papers*, 2000, pp. 156–157.

[41] Q. Huang and C. Menolfi, "A 200nV offset 6.5nV/$\sqrt{\text{Hz}}$ noise PSD 5.6kHz chopper instrumentation amplifier in 1μm digital CMOS," in *ISSCC Digest of Technical Papers*, 2001, pp. 362–363.

[42] J.L. Dawson and T.H. Lee, "Automatic phase alignment for high bandwidth Cartesian feedback power amplifiers," in *IEEE Radio and Wireless Conference*, 2000, pp. 71–74.

[43] J.L. Dawson and T.H. Lee, "Automatic phase alignment for a fully integrated CMOS Cartesian feedback power amplifier system," in *ISSCC Digest of Technical Papers*, 2003, pp. 262–263.

[44] S.M. McBeath, D.T. Pinckley, and J.R. Cruz, "W-CDMA power amplifier modeling," in *Proceedings of the IEEE Vehicular Technology Conference*, 2001, vol. 4, pp. 2243–2247.

[45] C.P. Silva, C.J. Clark, A. A. Moulthrop, and M.S. Muha, "Optimal-filter approach for nonlinear power amplifier modeling and characterization," in *IEEE MTT-S Symposium*, 2000, pp. 437–440.

[46] S. Watanabe, "Power amplifier modeling issues and advancement in Japan," in *Proceedings of the IEEE GaAs IC Symposium*, 1999, pp. 69–72.

[47] D. K. Choi and S. I. Long, "A physically based analytic model of FET class-E power amplifiers–designing for maximum PAE," *IEEE Transactions on Microwave Theory and Techniques*, vol. 47, no. 9, pp. 1712–1720, September 1999.

[48] C. Clark, G. Chrisikos, M. Muha, A. Moulthrop, and C. Silva, "Time-domain envelope measurement technique with application to wideband power amplifier modeling," *IEEE Transactions on Microwave Theory and Techniques*, vol. 46, no. 12, pp. 2531–2540, December 1998.

[49] M.Honkanen, O.Pöllänen, J. Tanskanen, E. Järvinen, and S. Häggman, "Comparison of measured and simulated $\frac{\pi}{4}$-DQPSK adjacent channel power using a functional high power amplifier model," in *Proceedings of the IEEE Vehicular Technology Conference*, May 1998, vol. 3, pp. 2459–2463.

[50] M. Honkanen and S. Häggman, "New aspects on nonlinear power amplifier modeling in radio communication system simulations," in *Proceedings of the 8th IEEE International Wireless Symposium on Personal, Indoor, and Wireless Communications*, 1998, vol. 3, pp. 844–848.

[51] R. Achatz, R. Dalke, and Y. Lo, "Power amplifier model for optimizing battery current, interference, and link margin," in *Proceedings of the IEEE International Symposium on Electromagnetic Compatibility*, 1998, pp. 434–438.

[52] A. Bernardini and S. De Fina, "Application of neural waveform predistortion to experimental TWT data," in *Proceedings of the 6th Mediterranean Electrotechnical Conference*, 1991, pp. 468–471.

[53] K. Nagaraj, J. Vlach, T. R. Viswanathan, and K. Singhal, "Switched-capacitor integrator with reduced sensitivity to amplifier gain," *Electronics Letters*, vol. 22, pp. 1103–1105, October 1986.

[54] P. Gray and R. Meyer, "MOS operational amplifier design–a tutorial overview," *IEEE Journal of Solid-State Circuits*, vol. SC-17, no. 6, pp. 969–982, December 1982.

[55] T.C. Choi, R. T. Kaneshiro, R. W. Brodersen, P. R. Gray, W. B. Jett, and M. Wilcox, "High-frequency CMOS switched-capacitor filters for communications application," *IEEE Journal of Solid-State Circuits*, vol. SC-18, pp. 652–664, December 1983.

[56] K. Bult and G. Geelen, "A fast-settling CMOS op-amp for SC circuits with 90-dB DC gain," *IEEE Journal of Solid-State Circuits*, vol. SC-25, pp. 1379–1384, December 1990.

[57] R. Gregorian and G. Temes, *Analog MOS Integrated Circuits for Signal Processing*, Wiley Series on Filters: Design, Manufacturing, and Applications. John Wiley and Sons, Inc., NY, NY, 1986.

[58] T.B. Cho and P.R. Gray, "A 10 b, 20 Msample/s, 35 mW pipeline A/D converter," *IEEE Journal of Solid-State Circuits*, vol. 30, pp. 166–172, March 1995.

[59] M. Pelgrom, A. Duinmaijer, and A. Welbers, "Matching properties of MOS transistors," *IEEE Journal of Solid-State Circuits*, vol. 24, pp. 1433–1440, 1989.

[60] J. Crols and M. Steyaert, "A 1.5 GHz highly linear CMOS downconversion mixer," *IEEE Journal of Solid-State Circuits*, vol. 30, pp. 736–742, July 1995.

[61] S. Galal, M. Tawfik, and H. Ragaie, "On the design and sensitivity of RC sequence asymmetric polyphase networks in RF integrated transceivers," in *Proceedings of the 1999 IEEE International Symposium on Circuits and Systems (ISCAS)*, July 1999, vol. 2, pp. 593–597.

[62] S. Galal, H. Ragaie, and M. Tawfik, "RC sequence asymmetric polyphase networks for RF integrated transceivers," *IEEE Transactions on Circuits and Systems–II: Analog and Digital Signal Processing*, vol. 47, pp. 18–27, January 2000.

Index

MIX
Papier aus verantwortungsvollen Quellen
Paper from responsible sources
FSC® C105338

If you have any concerns about our products,
you can contact us on
ProductSafety@springernature.com

In case Publisher is established outside the EU,
the EU authorized representative is:
Springer Nature Customer Service Center GmbH
Europaplatz 3, 69115 Heidelberg, Germany

Printed by Libri Plureos GmbH
in Hamburg, Germany